Elementary Mathematics from a Higher Standpoint

Felix Klein

Elementary Mathematics from a Higher Standpoint

Volume III: Precision Mathematics and Approximation Mathematics

Translated by Marta Menghini
in collaboration with Anna Baccaglini-Frank

Mathematical advisor for the English translation:
Gert Schubring

Springer

Felix Klein

Translated by: Marta Menghini
Mathematical advisor for the English translation: Gert Schubring

ISBN 978-3-662-49437-0 ISBN 978-3-662-49439-4 (eBook)
DOI 10.1007/978-3-662-49439-4
Library of Congress Control Number: 2016943431

Translation of the 3rd German edition „Elementarmathematik vom höheren Standpunkte aus“, vol. 3 by Felix Klein, Grundlehren der Mathematischen Wissenschaften in Einzeldarstellungen, Band 16, Verlag von Julius Springer, Berlin 1928.

Printed on acid-free paper

This Springer imprint is published by Springer Nature
The registered company is Springer-Verlag GmbH Berlin Heidelberg

Preface to the 2016 Edition

The present volume contains the translation of Felix Klein's "Präzisions- und Approximationsmathematik", which appeared in 1928 as volume III of the series *Elementarmathematik von einem höheren Standpunkte aus*[1] (Elementary mathematics from a higher standpoint). In turn, the 1928 edition was a re-edition of the lecture course delivered by Felix Klein in 1901 with the title "Anwendung der Differential- und Integralrechnung auf die Geometrie: eine Revision der Prinzipien", published in 1902 in lithographic form[2] (a reprint of 1908, edited by Conrad Heinrich Müller, left the text essentially unchanged).

Klein participated, together with Fritz Seyfarth, in the whole project of re-editing the three volumes on *Elementary mathematics from a higher standpoint*, but he died in 1925, after the first two volumes had appeared. The third volume was, therefore, edited only by Seyfarth; changes and insertions had been nevertheless discussed with Klein, as Seyfarth writes in his preface. The notes added are probably only by Seyfarth. We left, as in the 1928 edition, the notes by Seyfarth in square brackets, and labelled "translator's note" the notes added in the present edition. Of course, the notes by Klein himself are included without any additional marks.

In the text the reader will find, again in square brackets, *the page numbering of the original edition*. Cross references in notes and in the text *refer to this numbering*, as well as *the name index and the subject index* (that is, the original text has not been changed to this respect). Similarly for the references to Volume 1 and 2 of 1924/25, whose original layout is marked in the current new edition. Moreover, we left, as in the German edition, the comma to separate the integer part from the fractional part of a decimal number.

[1] Felix Klein, 1928. *Elementarmathematik vom höheren Standpunkte aus, 3: Präzisions- und Approximationsmathematik*; ausgearbeitet von C. H. Müller; für den Druck fertig gemacht und mit Zusätzen versehen von Fr. Seyfarth. 3. Aufl., Berlin: J. Springer, 1928 Series: Die Grundlehren der mathematischen Wissenschaften, 16. Reprint in 1968.

[2] Felix Klein, 1902. *Anwendung der Differential- und Integralrechnung auf die Geometrie: eine Revision der Prinzipien: Vorlesung gehalten während des Sommersemesters 1901*; ausgearbeitet von C. Müller, Leipzig: Teubner.

In the present translation we have added, when possible, the first names of the persons mentioned. In the German edition, as it was customary at that time, the first names were indicated only with the initials. The bibliographic references in the notes have also been completed, when needed.

"Präzisions- und Approximationsmathematik" was never translated into English, unlike the first two volumes; although the translators of the first two volumes were aware of the existence of volume III, they gave no reason for not having translated it. While the Spanish, the Russian and the Japanese translations also did not include volume III[3], there is a complete translation of all the three volumes into Chinese – first published in 1989 by Hubei Educational Press, Beijing, and published again by Chiu Chan Publishing (Republic of China, Taipei) in 1996.

On the occasion of the 13th International Congress on Mathematics Education (ICME-13, Hamburg 2016), the will of the *International Commission on Mathematical Instruction* (ICMI), together with Springer, is of closing this gap in the English version. We also proudly note that Felix Klein was the first president of ICMI, which was founded in Rome in 1908.

A translation after nearly a century, joined with a new edition of the two first volumes, might seem strange. Undoubtedly, this translation has an historical value, since Klein is one of the greatest mathematicians of history[4]. It also has a mathematical value – because of the interesting approaches and the links with the applications. But, above all, it has a didactical value, concerning the training of mathematics teachers. The volumes on *Elementary Mathematics from a higher standpoint* are still apt to this aim, for the simple reason that today's school mathematics is not very different from the school mathematics proposed at the time of Klein. Even if today we can find some snapshots of the mathematics of the 20th century (which is indeed too specialised) and also contents of probability and statistics, the remaining curricular contents have been developed well before the 20th century. The *higher mathematics* for these contents is mainly found in the period of the systematization of the mathematical theories, which corresponds to the period of Klein's writings. And, indeed, most of the contents of Klein's volumes are still offered in the many courses (which often bear the same name of *Elementary mathematics from a higher standpoint*) that all over the world are devoted to pre-service training of mathematics teachers.

For this third volume in particular, the motivations for the kind of study it proposes are given by Klein himself during his lifelong career. The third volume focuses on those properties that applied mathematicians take for granted when studying certain phenomena from a mathematical point of view. These properties must be seen as supplementary conditions (and constraints) to be required for the ideal objects of pure mathematics. However, in the meantime, these very proper-

---

[3] There is a recent translation of Volumes I and II into Portuguese and it might be extended to volume III.

[4] Books by Felix Klein are still in use today: Klein's *Nicht-euklidische Geometrie* has been re-edited in 2006, the English version of the *Lectures on the Ikosahedron* re-edited in 2007, and his *Development of mathematics in the nineteenth century* was translated in 1979.

ties prove to be the more intuitive ones. Therefore the comparison moves towards another field: it is a comparison between properties that can be considered only in the theoretical field of abstract mathematics and properties that can be grasped by intuition. Here the problem proves to become pertinent for mathematics teaching.

Klein had always been convinced of the importance of maintaining the link of mathematics with its applications and had manifested this interest ever since his initial work at the university of Erlangen, in 1872. The necessary link between pure and applied mathematics has been aptly characterised by Klein, in the 6th of his American Conferences (The Evanston Colloquium, 1894, p. 46):

> *I should lay particular stress on the heuristic value of the applied sciences as an aid to discovering new truths in mathematics. Thus I have shown (in my little book on Riemann's theories) that the Abelian integrals can best be understood and illustrated by considering electric currents on closed surfaces. In an analogous way, theorems concerning differential equations can be derived from the consideration of sound-vibrations; and so on.*

In this way Klein stresses the importance of the connection of mathematics with other disciplines that can foster intuition; and, in particular, the importance of maintaining this connection for (university) teaching:

> *It is not only admissible, but absolutely necessary, to be less abstract at the start, to have constant regard to the applications, and to refer to the refinements only gradually as the student becomes able to understand them. This is of course nothing but a universal pedagogical principle to be observed in all mathematical instruction* (ibid., 50).

Given Klein's intense plea for the applications, one should remark, furthermore, that he had not only alerted, already in the first volume, in the context of the emergence of set theory, against pushing a formalist program for the foundations of mathematics too far, but he also had taken up the issue again here in volume III, advising against searching for the new only for the sake of doing it:

> *Provided that a deep epistemological need exists, which will be satisfied by the study of a new problem, then it is justified to study it; but if one does it only to do something new, then the extension is not desirable* (p. [157]).

In the preface of the 1928 edition of *Präzisions- und Approximationsmathematik* we do not find explanations of the fact that this lecture course, conceived with the purpose of filling the gap that was created between pure mathematicians and applied mathematicians, had become a teacher training course within the series of *Elementary mathematics from a higher standpoint*. Indeed, the re-editing of this lecture course was completed after Klein's death.

In any case, a reference to teaching in schools can be found in the preface to Volume III of 1902, but we find nothing explicit concerning Klein's general vision on applications. Klein only claims that teaching – for beginners as well as for those students who want to use mathematics only as an auxiliary means – has to make a naïve use of the intuitive approaches.

In place of Klein, we can quote the Italian geometer Guido Castelnuovo, who was very close to Klein's points of view. In 1913/14 at the University of Rome Castelnuovo used Klein's 1902 lecture course, recognising its significance for the students who were to become high school teachers. At the beginning of his lectures Castelnuovo asserted that the teaching and learning of mathematics would be more successful if it included, besides the logical procedures that lead to the theorems, also the way in which concepts are formed starting from observations, and the way in which these same theorems can be verified in practice. Castelnuovo therefore sees the necessity to distinguish, as does Felix Klein, between precision mathematics and approximation mathematics: the first one includes all the propositions that can be logically deduced from the axioms of geometry or of analysis – obtained by abstraction from experience; the second one includes the results that can be obtained from experience with a certain degree of approximation.[5]

The school that Castelnuovo had in mind was in particular the *modern lyceé*, which – in the wake of the reform movements of many European countries – had to include new topics, in particular differential and integral calculus, with wider application in the life sciences. Mathematics was there presented as an appropriate language for describing natural phenomena in which the concept of function assumes particular importance.

Still today, this kind of high schools that devote a greater number of hours to mathematics and sciences is widespread, and it is considered important to maintain the link between the different scientific fields and to show how mathematics arises from observation.

Maybe some of the examples offered by Klein will seem "old", like for instance the harmonic analyser for the calculation of the Fourier coefficients. In any case the functioning of this instrument can represent an historical means to clarify certain concepts. The motivations for proposing this kind of study are still, and possibly are even more so today, those given by Klein.

On the other hand, even though he starts from an intuitive and sometimes practical approach, Klein develops more abstract concepts working in rich "mathematical environments", which form the core of a pertinent program for mathematics teacher education at secondary schools.

We thank Leo Rogers for his careful re-reading of the book, and the many colleagues to whom we have asked advice, in particular Geoffrey Howson and Claudio Bernardi.

Marta Menghini and Gert Schubring

[5] See Menghini, Marta (2016), Precision and approximation mathematics for teacher education: the lecture course of Guido Castelnuovo and the influence of Felix Klein, in *Dig where you stand 4* (Bjarnadottir, Furinghetti, Menghini, Prytz and Schubring, eds), Proceedings of the 4th International Conference on the History of Mathematics Education (ICHME4), Turin, 23–26 September, 2015, Rome: Edizioni Nuova Cultura.

Preface to the First Edition

In publishing in lithographic form the lecture course that I gave last summer, I would like to take a new step in the main direction of my efforts over the last years: namely to give back to mathematical science – seen as a connected whole – its value with respect to all its branches, as it was taken for granted before there were specializations. But, in particular, once again to achieve a better mutual understanding among the representatives of abstract and applied mathematics, compared to that which presently exists. I have often emphasised that, in this respect, it would be of crucial importance to have a clear grasp of the differences and also of the mutual relations between the two parts of mathematics that I shall call *precision mathematics* and *approximation mathematics*. Similar thoughts have been expressed in recent years by Messrs. Heinrich Burkhardt and Karl Heun; see for instance the inaugural speech of the former (Zürich, 1897[6]), which recently became accessible to a wider circle of colleagues thanks to its reprint in the first issue of vol. 11 of the *Jahresbericht der deutschen Mathematiker-Vereinigung*, and the report of Mr. Heun on the kinetic problems in technology in Vol. 9 of the same journal (1900–1901). As far as I know, the term "Approximation mathematics" appears for the first time in this report; in any case I took it from there. A detailed explanation of the contrasts to be considered, like those presented for the field of geometry in the following lecture course, should thus be better understood and appreciated by a wider circle of people.

At the same time, with this lecture course, I complete those discussions that I often presented on the methods of mathematics teaching, particularly in higher education (see for instance *Jahresbericht*, vol. 8, 1898–1899). My opinion is still that the teaching for beginners and for those students who want to use mathematics only as an auxiliary means for other studies has to make a naïve use of the intuitive approaches; the conviction that this is necessary for pedagogic reasons, considering the disposition of the majority of students, has become noticeably stronger in the last few years, here and abroad.

[6] Mathematisches und naturwissenschaftliches Denken.

But I am no less convinced (something I never omitted to state) that – in accordance with today's development of science – such a form of teaching will not suffice for the training of the professional mathematician in high school; rather, that here must prevail, in addition to the facts of intuition, the central importance of the modern concept of number and of the far-reaching related developments. What I miss now, in the textbooks and the lecture courses that I have encountered, is the transition from one conception to the other. In this sense the following presentation needs to be inserted as a complement; its major aim will be reached if one day it will prove to be superfluous, because in this case the considerations, which it offers, will have become an obvious part of teaching in all the branches of higher mathematics.

Göttingen, February 28, 1902.

Felix Klein

Preface to the Second Edition

The following is essentially an unchanged reprint of the first edition of the lecture notes in 1902. Only some inaccuracies have been amended, and occasional references to recent publications that are intimately connected to the topics developed in the lecture have been added. At the end, the "Report of the Göttingen Faculty of Philosophy concerning the task for the Beneke prize 1901" is added, to which different remarks in the text refer.

Göttingen, January 5, 1907.

Conrad Heinrich Müller

Preface to the Third Edition

The lecture notes of F. *Klein*, which are now appearing as volume III of "Elementary mathematics from a higher standpoint"[7] with the subtitle "Precision and Approximation Mathematics", were written some years before the lecture courses to be found in the two first volumes. Like these latter, this course is intended for a wide audience because of its goals and the manner of presentation. The lecture notes, in their lithographic form, have for a long time been cited in mathematical literature with the title "applications of differential and integral calculus to geometry (a revision of the principles)". The change of title is due to the personal request of *Felix Klein*, with whom I had – in the last two months before his death – a series of conversations about the work required for their publication. Klein believed that the new title would better meet the goals of the notes than the former.

In editing the lecture notes, the same principles, which have inspired the publication of the first two volumes, have been followed. The original presentation was essentially maintained, but single changes and insertions were added when practical and stylistic considerations were considered necessary. The additions, some of which were due to the need to take into account more recent publications, were inserted in the text when this was required by the organization of the lecture course, otherwise they were added in the form of footnotes. The figures were often improved and expanded, particularly in the final part of the lecture, which concerns gestalt relations of curves in space and surfaces of the third degree. This section, which shows *Klein* to be the master of the art of impressive representation of geometric forms, as he is widely recognised, was originally conceived only for those readers for whom the discussed models were available. Here, it was necessary to structure the text in such a way, that the reader needs no longer to refer to these models, by adding figures and explanations. For the additions, I made much use of a lecture by *Klein* in 1907 about the theory of surfaces and spatial curves. The

[7] Translator's note: The German title is *Elementarmathematik vom höheren Standpunkte aus. Dritter Band: Präzisions- und Approximationsmathematik.*

"report of the Göttingen University Faculty of Philosophy concerning the task for the Beneke prize 1901", which was added to the second edition of 1907, is now omitted, as it has already been reprinted in volume II of *Klein*'s Gesammelte mathematische Abhandlungen.

In the edition of the text I was assisted by Mr. *Alwin Walther*, Göttingen, who made many valuable suggestions. Also Mr. *Hermann Vermeil* contributed to the reading of the proofs and he prepared the new indexes. My colleague *H. Homann* assisted me by photographing a series of models in the mathematical collections in Göttingen. With Prof. *Richard Courant*, in particular, I had some conversations that were important for the progress of my work. I owe all the above persons my hearty thanks. I am also very grateful to the publishing company Julius Springer for the willingness with which they met my requests.

Göttingen, January 31, 1928.

Fritz Seyfarth

Contents

Introduction

Modern mathematical literature is characterized by a far-reaching distinction, which all of you must have noticed: The interests and ways of thinking of the theoreticians differ considerably from the methods that are effectively used in mathematical applications. Not only does the scientific training of the individual suffer from this discrepancy, but also the standing of this science itself. It seems extremely important to try to counteract the grievances arising from this situation. The lecture course that I start here is aimed at providing a contribution to this issue, in that it tries to establish a clear relationship – so to say from an epistemological point of view – between the different kinds of mathematical questions, as they arise in some occasions according to their very nature. You must, on the one hand, try to understand the interests of modern theoreticians, but on the other hand you must form an opinion on which parts of mathematical speculation have a direct meaning in applications. I do not doubt that you will find the resulting contraposition of different points of view interesting and beneficial. My hope is that this lecture course will eventually even result in you, on your part, contributing in the future to re-shape the considerably single-sided development of our science towards an all-round, harmonious one!

However, the program that I am setting up is too broad to be developed in all directions in one semester. Therefore I will bring to the foreground only *one* branch of mathematics, namely *geometry*. The practical part of geometry includes geometrical drawing and measurement, which is confronted – from a theoretical point of view – with the abstract treatment of geometrical problems started by the Greeks. I will treat the questions relevant for both sides in analytic form. This is actually not necessary; one could also proceed without leaving the field of geometry. However, the development of the questions that I have to discuss has, until now, mainly taken place in connection with *analysis*: We will consider in particular the development and meaning of differential and integral calculus.

I cannot give any textbook as reference. I will give single references case-by-case and for the rest I prefer to refer to the *Enzyklopädie der mathematischen* **[2]** *Wissenschaften*, which aims to present all the mathematics of the 19th century and to arrange all its publications.

F. Klein, *Elementary Mathematics from a Higher Standpoint*,
DOI 10.1007/978-3-662-49439-4_1

If one wants to extend the same programme to other branches of mathematics, this would particularly concern the issues of the mathematical explanation of nature (mechanics, etc.). A series of pertinent ideas are contained in the Report elaborated by our Faculty of Philosophy in the year 1901 on the occasion of the celebration of the Beneke-Stiftung[8]. For the rest, I will refer to a paper of my own, which I published in the *Erlanger Berichten* in 1873 with the title "Über den allgemeinen Funktionsbegriff und dessen Darstellung durch eine willkürliche Kurve"[9]. I can also refer to the sixth lecture of my Evanston Colloquium "Lectures on Mathematics", New York 1894[10], translated into French by Léonce Laugel, Paris 1898. The title of that lecture is "On the mathematical character of space intuition and the relation of pure mathematics to the applied sciences".

Now I immediately start with the first part of my lecture course:

[8] [Printed in *F. Klein*: Gesammelte mathematische Abhandlungen, Vol. II, pp. 241–246]. Translator's note: *Felix Klein*, Auszug aus dem Gutachten der Göttinger philosophischen Fakultät betreffend die Beneke-Preisaufgabe für 1901, *Mathematische Annalen*, Vol. 55 (1901, 143–146). *Friedrich Eduard Beneke* (1798–1854) was a philosopher and psychologist at Berlin and Göttingen universities. He established a price award which became regulated by the Göttingen Königliche Gesellschaft der Wissenschaften in 1867, demanding the Philosophical Faculty of Göttingen University to publish yearly a subject for a price essay. The subject of 1898 for 1901 was to question the principles of continuity and the use of unrestrictedly differentiable functions for the study of the natural phenomena and to unravel in particular preconditions having remained implicit, in connection with limitations of sensory perception.

[9] Reprinted in *Mathematische Annalen*, Vol. 22 (1883) [and in *F. Klein*, *ibid.*, pp. 214–224]

[10] [New edition, obtained from the American Mathematical Society. New York 1911. The mentioned talk can be found also in *F. Klein*, *ibid.*, pp. 225–231]

First Part: On Functions of a Real Variable and Their Representation in an Orthogonal Coordinate System

I. Remarks on the Single Independent Variable x

Empirical and Abstract Precision. The Modern Concept of Number

I am beginning systematically in such a way that I will not speak of functions at all, but instead introduce the *independent variable* x and represent it in the usual manner as a point on the x-axis[11] whose distance from the origin is $|x|$ units of measurement and which lies to the right of the origin for positive values of x and to the left of the origin for negative ones. I wish to direct your attention to the *precision* of the construction of such a point. I note in this respect: If I use an empirical procedure such as drawing, measuring, evaluating by eye, or mental reproduction by spatial **[3]**
perception, the result of the construction only has limited accuracy; I will prove this for measuring, by showing some numerical values.

If we use, in measuring, the greatest precision that we can reach today with the help of a microscope and a screw thread micrometer, we cannot reach a higher precision than 0,1 μm per metre. If we want to go further, intensifying the microscope, light could fail to have any effect because of diffraction and the molecular structure of the matter would react to it; thus the physical laws to which light and matter are subjected seem to prohibit a major precision in measurement. Expressed in metres, 0,1 μm is equal to 10^{-7} m, so that *the lengths in metres found by means of direct measuring can be reliable at most to the 7th decimal.*

The situation is similar when drawing or measuring by eye, only that in this case the threshold, that is the value beyond which the precision of the observation cannot increase, is of course much higher.

As a major theorem, which is not a mathematical theorem, but which is fundamental for all applications of mathematics, we can state that: *In all practical fields there exists a threshold of precision.*

What about spatial perception? With this question we touch a controversial field, in which philosophers have presented a great variety of views through history. In particular, for many philosophers spatial perception appears as something absolutely precise[12]. On the contrary, we have to underline that the modern math-

[11] Translator's note: The literal translation is *axis of abscissas*; see also note 23.

[12] Translator's note: Given the crucial importance of "Anschauung" in Klein's epistemology of mathematics, one might think that he agrees with Kant's conception of "Anschauung", which is

F. Klein, *Elementary Mathematics from a Higher Standpoint*,
DOI 10.1007/978-3-662-49439-4_2

ematician can give many examples of spatial structures where spatial perception appears to be absolutely impossible because of the fineness of the structure considered. Thus, regarding spatial perception, I believe that there also exists a threshold; I do not want to impose this opinion upon you, but I will point now and again to the structures mentioned leaving it to you to check their spatial imaginability. I state the following assertion referred to spatial perception: *The existence of a threshold in perceiving spatial structures is controversial and can be left to repeated checking in the examples that will be presented.*

Now, let us compare these empirical procedures for the determination of the number x with the *arithmetic definition* of such number, as it is given in the theory of real numbers. You will observe that in the arithmetic definition *precision has no limits*. Let us first agree on the arithmetic definition of a real number x, in short, on the *modern concept of number*.

[4] In order not to have to go too far back, we will say that we understand the totality of the real numbers to be represented by the totality of the decimal fractions[13], where we also consider an *infinite* decimal fraction as representing a specific number. Vice versa, will every real number be represented by a single specific decimal fraction?

We introduce here the following theorem[14]: We can modify every finite decimal fraction by decreasing the last digit by one unit and having nines in all the following positions. But this is, as one can prove, the only indetermination that we meet when representing real numbers by means of decimal fractions. Therefore: *Every finite or infinite decimal fraction yields a number, and every number, with the exception of the above mentioned case, only yields one decimal fraction.*

What is the difference, in this representation, between rational and irrational numbers? The answer is known from the Elements: Written as a decimal fraction, a rational number $\frac{m}{n}$, with m and n integer numbers $\neq 0$, is characterized by the fact that the decimal fraction either terminates or continues periodically. The inverse is also true, as I mention incidentally.

Of course it is necessary to extend the usual computational rules to irrational numbers, since such rules are initially set up only for rational numbers. We suggest the comprehensible yet very thorough presentation contained in *Konrad Knopps'* Theorie und Anwendung der unendlichen Reihen (2nd edition, Berlin 1924), which proposes this classification of irrational numbers. Moreover, you can

just as important for this philosopher. However, the first edition of 1902 contains here a rather distanced reference to Kant: "Es ist insbesondere *I. Kant* und seine Schule, die mit sehr bestimmten Auffassungen hervorgetreten sind. Es erscheint bei *Kant* die räumliche Vorstellung als etwas absolut Exaktes (nichts Verschwommenes)": It is in particular *Immanuel Kant* and his school who are known for very definite conceptions. Kant conceives spatial perception as something absolutely exact (not vague).

[13] Translator's note: In German "Dezimalbruch"; the concepts of *decimal fraction* and of *dotted decimal* are equivalent, but the former explicitly refers to the representation as a fraction with a power of 10 as denominator; see also volume 1.

[14] See Vol. I, p. [34]–[37]. Translator's note: The references are to the pages of the edition of 1924, see preface to the 2016 edition.

compare the report of *Alfred Pringsheim*: Irrationalzahlen und Konvergenz unendlicher Prozesse, vol. I, pp. 49–146 in the Enzyklopädie der mathematischen Wissenschaften (priority should be given to pp. 49 to 58[15]); these should be read by every mathematician.

Precision and Approximation Mathematics, also in Pure Geometry

And now we come to a point that I consider especially important: The representation of a real number by means of a decimal fraction is absolutely precise. The decimal fraction corresponds exactly to the real number; in abstract arithmetic, the threshold we mentioned before decreases below any limit. So we have the *fundamental difference between empiricism and idealization*, which in this particular case I formulate as follows: *In the ideal field of arithmetic there exists no threshold different from zero, contrarily to what happens in the empirical field; instead, the precision with which the numbers are defined or are considered to be defined is indefinitely high.*

The difference – seen here in the particular case of the empirical determination of a magnitude in practical geometry and its exact definition in abstract arithmetic **[5]** – between limited and unlimited precision can be found whenever we compare an external perception or a practical action with abstract mathematics. This difference exists for *time*, for *all physical or mechanical magnitudes*, and particularly for *numerical calculations*. For, what is calculation with seven-digit logarithms other than a calculation with approximations whose precision reaches only the seventh digit? On the other hand, we will see that in every field we can proceed to absolute exactness by means of suitable axioms; then we substitute an empirical structure with an idealized mental object.

This differentiation between absolute and limited precision will be the red thread of the whole lecture; it requires a *bisection of all mathematics*. We distinguish:

1) *Precision mathematics* (calculation with the real numbers),

2) *Approximation mathematics* (calculation with approximated values).

The term approximation mathematics does not represent a downgrading of this branch of mathematics, because it does not correspond to a rough mathematics, but to a precise mathematics of rough relations. We get the *whole* science only when we embrace the two parts:

Approximation mathematics is that part of our science that is effectively used in applications; precision mathematics is, so to speak, the solid structure on which approximation mathematics is grounded.

First of all, I want to describe this bisection of mathematics with the example of *numerical calculations*, where the contrast mentioned appears immediately and

[15] [The paper was finished in 1898. See also *Alfred Pringsheim:* Vorlesungen über Zahlenlehre, Erste Abteilung. Leipzig 1916]

is not obscured by complicated psychological arguments. In precision mathematics we work with real numbers, as they have been represented before by means of finite or infinite decimal fractions. Now, how can one indicate, for instance, in the language of precision mathematics, the approximate value of the number $x = 6{,}4375284\ldots$, which is given exactly up to 7 decimals? Mathematics has possessed, ever since the 18th century, the functional symbol *E(x)*[16], a shortcut for the French term "entier". $E(x)$ is the greatest integer contained in x. With the help of this symbol the previous approximated value can be expressed as $\frac{E(10^7x)}{10^7}$, so that – when calculating with seven-digit decimals – we operate with the number $\frac{E(10^7x)}{10^7}$ in place of the number x[17].

[6] If we take the conceptualization of precision mathematics as our gauge, we can say: *In approximation mathematics we do not work with the numbers x, but with the E-functions of the numbers x.*

Similarly we can formulate the *difference between empirical and abstract geometry*, or, better, the relation of empirical geometry to abstract geometry.

First, let us speak of the nature of *practical geometry*. It is the kind of geometry with which we have to deal with if we want to master concrete spatial situations, for instance when drawing or modelling, or in land surveying. Here all the operations are exact only up to a certain threshold. Therefore for these disciplines we have to formulate definitions and principles of the following type:

A point, in these disciplines, is a solid that has such small dimensions that we can disregard them.

A curve, in particular a straight line, is a sort of stripe in which the width is negligible if compared to the length.

Two points determine a straight line connecting them, which is the more accurate the further they are apart. If they are very close, they are quite unsuitable for determining a straight line.

Two straight lines determine an intersection point in a way that is more precise, the less the difference is between the angle that they form and a right angle. When the angle becomes smaller and smaller, the inaccuracy of their intersection point increases.

But now let us speak of *abstract geometry*: here we have no such indefinite statements. This depends on the fact that abstract geometry starts by putting up front, in the form of axioms, those statements that are given in an approximate formulation in practical geometry. So here we are at a very crucial conceptual turning point. Relations, which are, in practice, only approximately correct, are postulated as exact, and – in abstract geometry – logical conclusions are drawn on the basis of the agreed axioms. The formulations corresponding to the preceding ones are now the following:

---

16 Translator's note: Today the floor function or integer function is denoted by $y = [x]$.

17 If we do not truncate the decimal fraction, but round it, everything remains valid as long as we define $E(10^7x)$ as the greatest integer contained in the interval from $10^7x - \frac{1}{2}$ to $10^7x + \frac{1}{2}$.

A point has no spatial extension.

A line has only length.

Two points determine a straight line in an exact way.

Two straight lines that meet have a precise intersection point.

I now want to draw your attention especially to the *link between these two completely different kinds of geometry* (corresponding to the transition that just occurred in arithmetic through the introduction of the E-function). Everybody tacitly admits that such a link exists. The results of abstract geometry are always applied to prac- **[7]**
tical situations, even though the foundations of abstract geometry go beyond the constraints of practical geometry. If this must happen in a rational way, it is necessary to develop what is called the *theory of observation errors.* We say: Each observation or practical construction is subject to an error. It would be more correct to say: Each observation does not establish – in the sense of abstract geometry – a specific magnitude, but rather *intervals for each quantity.* And the question, 'In what form does abstract geometry find its applications in practical geometry?' can then be answered in that way: *We have to investigate, within what limits the result of a mathematical evaluation changes if we consider the data that are at the base of such evaluation as varying in an interval.*

Now I am going back to the starting point of our reflections, when we associated numbers with the points of an axis and added that such an operation could be performed in practice only with limited precision. If we want to have a precise conformance, we have to determine it by means of an axiom, which cannot be checked empirically, in this way we enter the realm of abstract geometry.

We postulate that each real number x (in the terms of the above definition) corresponds to one and only one point of the x-axis, and also that each point x corresponds to one and only one number. This axiom is the basis of analytic geometry – or we could concisely say of *geometry of precision,* if we want to give to it a precision-theoretic character[18]. Now, I do not deny that intuition *drives* us to such a definition, but I maintain that with such a definition we go *beyond our intuition and our imagination.* I can clarify what I mean with the help of two examples taken from the realm of set theory, which is playing a relevant role in today's mathematics.

A set of numbers, and therefore – according to our definition – also a point set on the x-axis, is a collection of numbers or points x, defined in such a way that for every number or point x it is certain whether such point belongs to the collection or not[19].

Let us take, for example, two simple sets of points:

a) *All rational numbers or points* $x = \frac{m}{n}$ *from 0 to 1.*

These are "everywhere dense" on the interval $0 \dots 1$. If you take a subinterval, **[8]**
however small, it still contains infinitely many rational numbers and yet this subinterval – if such a rough language is allowed – is still riddled like a sieve, as infinitely

[18] [The above-mentioned Axiom is known as the *Cantor–Dedekind axiom* of continuity.]

[19] Translator's note: Actually, the sets that satisfy this property are called *decidable sets.* Klein will later refer to them as *well-defined sets.*

many irrational numbers, which do not belong to our point set according to the definition, have to find a place in it. The question is now: Does our spatial perception allow us to imagine this, a dense point set that does not exhaust the continuum? The abstract definition is clear, the representation is, in my opinion, impossible.

b) *All numbers between 0 and 1 with the exclusion of the endpoints 0 and 1*, this means an open interval.

Our spatial perception seems to teach us that we cannot truncate a segment without considering its endpoints. Therefore we can hardly represent spatially the collection of all numbers between 0 and 1 with the exclusion of the two endpoints, even if this certainly identifies a well-defined set.

Intuition and Thought, Explained by Different Examples from Geometry

These remarks have an *important methodological consequence*, which I formulate as follows:

As the objects of abstract geometry cannot be totally grasped by space intuition, a rigorous proof in abstract geometry can never be based only on intuition, but it must be founded on logical deduction from valid and precise axioms[20]. Nevertheless *intuition* maintains, also in precision geometry, its irreplaceable value that cannot be substituted by logical considerations. Intuition helps us to construct a proof and to gain an overview, it is, moreover, a source of inventions and new mental connections.

You can compare, in this respect, the inaugural lecture:

Otto Hölder: Anschauung und Denken in der Geometrie. Leipzig 1900.

I highlighted earlier what the differences between precision mathematics and approximation mathematics are, and that only the two together can build the whole science. The mistake from which todays' science suffers is that the theoreticians are concerned too unilaterally with precision mathematics, while the practitioners use a sort of approximate mathematics, without being in touch with precision mathematics through which they could reach a real approximation mathematics. I want to clarify this contrast between theoreticians and practitioners with a further example, which I take from the usual teaching of elementary geometry.

[9] Ever since Euclid, elementary geometry has been mainly considered as precision mathematics; certain axioms (that are postulated to be exactly valid) are placed at the beginning, from which new theorems and proofs are deduced in a purely logical way. As a consequence, distinctions that are meaningless for practical geometry are needed.

In this respect I first mention the *difference between commensurable and incommensurable segments*. Please note that if we only use limited precision this

[20] Probably any theoretical mathematician today will agree with this requirement; but it seems convincing to me only if it is based, as in the text, on the inaccuracy of our spatial perception.

distinction makes no sense whatsoever. For instance, the diagonal of the unit square is $\sqrt{2} = 1,414\ldots$. We can denote this decimal fraction with increasing approximation by $\frac{14}{10}, \frac{141}{10^2}, \frac{1414}{10^3}$ that is, we can approximate it close enough, as required by an ever lower threshold. If we pose the threshold, for instance, equal to $\frac{1}{10^6}$, then any practical needs of measuring geometry are met. To go beyond this has only a theoretical meaning. Therefore:

The difference between commensurable and incommensurable in its strict sense (and hence also the concept of irrational number) belongs solely to precision mathematics.

But in what sense will the practitioner use the terms "commensurable" or "incommensurable", if he does not want to banish them from his vocabulary? I am thinking now of astronomy and in particular of perturbation theory. Take for instance the influence of Jupiter on a small planet, like Pallas. What does the astronomer mean when he calls the ratio of the orbital periods of Pallas and Jupiter commensurable? The period can be determined only with limited precision, due to the various kinds of perturbations, which have to be considered; that is – in the usual units – it can be expressed exactly only up to a certain number of digits. Therefore, if the astronomer says, 'Two planets have commensurable orbital periods', he can only mean that the periods behave like two small integers, in the case of Jupiter and Pallas like 18 : 7. But if the ratio found through observation were, for instance, 180000 : 73271, one would no longer call the periods commensurable.

Let us now go back to Euclid and ask ourselves the question: Is a certain problem solvable by ruler and compass, that is, does a certain number of repeated applications of those instruments lead to a solution? Here, too, we have to give the following answer:

The question as to whether a segment can be constructed with ruler and compass **[10]** *also belongs to the realm of precision mathematics; it refers to the existence of an exact solution under the given axioms.*

In this regard, Antiquity has left us, as is well known, three fundamental problems that were left as probably unsolvable, whose effective insolubility was in fact later proved in modern times. They are:

a) *The trisection of an arbitrary angle*

b) *The duplication of the cube, that is the construction of* $\sqrt[3]{2}$,

c) *Squaring the circle, that is the construction of the number* $\pi = 3{,}1415926\ldots$ *starting from the unit segment.*

I now want to draw attention to the fact that the assertion: It is impossible to solve the problems a) to c) through a finite number of applications of ruler and compass, belongs to precision mathematics and not to approximation mathematics. From a practical point of view you can solve each of the problems exactly, because it is empirically possible to do it with ruler and compass[21]. In this manner, for instance, π can be easily approximated by the construction of the polygons in-

[21] [On the related proofs of impossibility compare Vol. I, pp. [56]–[60], pp. [123]–[124], pp. [256]–[269]. For the approximation method you can refer in particular to the book by *Theodor Vahlen*: Konstruktionen und Approximationen (Leipzig, 1911)]

scribed and circumscribed to a circle. But also far simpler and often very ingenious approximations have been found. In many cases a purely practical method is suitable. To divide an angle in three equal parts we can keep trying until the division is sufficiently accurate; this method is, in practice, much more efficient than calculating it up to a certain number of decimals and reproducing then the value found in the figure (compare the circular dividing engines for geodetic and astronomical instruments[22]).

Explanation Through Two Simple Propositions on Point Sets

Now I go back again to our reflections concerning point sets. We had reached the conclusion that in proofs within precision mathematics we could not rely on intuition, as significant as any drawing may be to get an idea about hypotheses for the respective proof. Now I give two very simple theorems on point sets where this already proves to be pertinent; it will be important to understand what the proofs effectively achieve.

1. On the x-axis an infinite point set is given, which does not pass a given fixed
[11] point A (it does not even need to come close to A). We say that such a point set is *bounded from above (or from the right)*. Then we claim that: *There is an upper (right) limit of the set.*

2. On the x-axis an infinite point set is given which lies between two points A and B. *Then, between A and B there exists at least one accumulation point, that is a point such that in any however small neighbourhood of it lie an infinite number of points of the set.*

When we hear such theorems for the first time, they can seem obvious. Of course, they are obvious, but in a higher sense that sums up all mathematical insight, namely in the sense that mathematics is basically the science of the obvious things. We can and want to prove nothing surprising with our theorems, we are only trying to show in a clear way how the theorems are connected to the concept of number, which we have put at the foreground of our developments.

What about our first theorem?

The point set is well-defined, that is, we can establish for each point whether it belongs to the set or not. Now it can happen that the set includes a point that has effectively the greatest abscissa[23]; then it surely represents the upper limit we were looking for. Examples for such sets are: a) the set of all points between 0 and 1, with the inclusion of 1; b) the set which includes all the points of the previous set with the addition of the point 2. In the first case the upper limit is 1, in the second case 2. But the set can be such that it does not contain a point which is the farthest

[22] [More details can be found in *Leopold Ambronn*: Handbuch der astronomischen Instrumentenkunde Vol. I (Berlin 1899), Cap. IX, pp. 433–437.]

[23] Translator's note: Abscissa and ordinate are nowadays usually referred to as x-coordinate and y-coordinate. However the terms abscissa and ordinate also have the geometrical meaning of *segments* measuring the distance from a point to the y-axis or the x-axis resp.

to the right, that is, it represents – if we think of its points as being ordered – a monotonically increasing set; although the set does not pass a given point A on the x-axis, it has no maximum. An example of this kind is the set that we have seen before, the set of all points between 0 and 1, with the exclusion of the endpoints. In this case we will mean by upper limit a point which does not belong to the set, but that to its left is preceded by points of the set that are arbitrarily close. We surely cover all three cases of the given examples when we define:

The point G, not necessarily belonging to the considered set, is called the upper limit of the set if to the right of G there is no point of the set, and if to the right of $G - \varepsilon$, however small ε may be, we can find always at least one point of the set.[24]

Once we have agreed on this, we turn to the representation of a real number **[12]**
through a decimal fraction with infinitely many decimals and prove the existence of the upper limit in this way:

First we look at which integers are exceeded by points of the set and which are not. This is possible because we consider the set as well-defined. If a is the greatest number that is exceeded, then $a + 1$ is not. We now have two possibilities. Either $a + 1$ is a point of the set, or it is not. In the first case, $a + 1$ is the upper limit of the set. In the second case let us divide the interval $a \ldots a + 1$ in 10 equal parts, so we obtain, with an easily understandable abbreviation, the following decimal fractions $a = a{,}0$; $a + 0{,}1 = a{,}1$; $a + 0{,}2 = a{,}2$; $\ldots$; $a + 0{,}9 = a{,}9$; $a + 1 = a + 1{,}0$.

We compare these numbers with the points of the set and we find again a greatest number a, a_1, which is again exceeded by an element of the set, while $a, a_1 + 1 (a + 1{,}0$ in the case $a_1 = 9)$ is either reached or not reached.

In the second case we divide the interval $a, a_1 \ldots a, a_1 + 1$ again in 10 equal parts and we have as endpoints

$$a, a_1\, 0; \quad a, a_1\, 1 \ldots; \quad a, a_1\, 9; \quad a, a_1 + 1{,}0.$$

We compare these numbers again with the points of the set and obtain two numbers $a,\ a_1\, a_2$ and $a,\ a_1\, a_2 + 1$. We can see how the process goes on and how it either yields a number, after a finite number of steps, that belongs to the set and represents the upper limit we were looking for, or it leads to an infinite sequence of intervals, each of which is contained in all the preceding ones, and whose length decreases to zero. Due to Weierstraß's definition of the equality of real numbers, the two sequences of the left and right endpoints of the interval determine the same decimal fraction[25]. This then represents the upper limit in question. I briefly summarize the idea:

We prove the existence of the upper limit by indicating a way in which we can produce such a limit as an exact decimal fraction.[26]

---

[24] In the case in which G does not belong to the set, we deduce from the above definition – as ε can be chosen arbitrarily small – that in the interval $G - \varepsilon < x < G$ lie an infinite number of points of the set. Then G is an accumulation point of the set. However, if G belongs to the set this is not necessary, as shown in the example 2 (isolated upper limit).

[25] [See Vol. I, pp. [36]–[37]]

[26] [In an analogous way we can prove the existence of the lower limit]

The generality of the second theorem, which *Karl Weierstraß* particularly emphasized in his lectures for its importance in the theory of functions, is not affected if we take as endpoints of the interval, within which the infinite point set must have at least an accumulation point, the two consecutive integers a and $a+1$.

[13] Denoting the accumulation point by x_0, its definition is:

x_0 is an accumulation point if between $x_0 - \varepsilon$ and $x_0 + \varepsilon$, however small we choose our strictly positive ε, there are infinitely many points of the set.

We give the proof of our theorem again in such a way that we actually construct a decimal fraction x_0 with the desired property, or rather we think about how we could concretely construct it.

At first, we divide the interval between a and $a+1$ again in 10 equal parts

$$a{,}0 \quad a{,}1 \quad a{,}2 \ldots a{,}9 \quad a+1{,}0$$

and acknowledge that at least in one of these intervals there are infinitely many points of our infinite point set.

Let this interval be $a{,}a_1 \quad a{,}a_1+1$. For this interval, whose length is only $\frac{1}{10}$ of the initial one, we repeat the argument after having divided it into 10 equal parts

$$a{,}a_1 0 \quad a{,}a_1 1 \ldots a{,}a_1 9 \quad a{,}a_1 + 1{,}0$$

Let the interval obtained in this manner (its length is $\frac{1}{100}$ of the initial one), in which there still are infinitely many points of our set, be $a{,}a_1a_2 \quad a{,}a_1a_2+1$. You can see how we again obtain an infinite sequence of intervals, each of which is contained in all the preceding ones and whose length tends to zero. The infinite decimal fraction x_0 determined by the interval sequence represents now, in fact, an accumulation point for the given point set, since in each interval of the sequence there are infinitely many points of the set. You can also observe here that: *The existence of an accumulation point x_0 is shown in such a way that we obtain the decimal fraction x_0 by means of a process following definite rules, which brings us back to our basic definition of the number as a decimal fraction with all its digits defined.*

The two theorems mentioned about point sets should be an example of the objects considered by precision mathematics even when treating a single independent variable, and of how the proofs have to be conducted – by virtue of the principle of absolute precision – going beyond intuition.

II. Functions $f(x)$ of a Real Variable x

The Abstract and the Empirical Definition of a Function (Idea of the Function Stripe)

So far, we have only referred to the definition of the independent variable x. Now we turn to the *functions* $y = f(x)$ asking ourselves what the questions concerning **[14]** precision mathematics and approximation mathematics might be here.

The most general definition of a function that we have reached in modern mathematics starts by fixing the values that the independent variable x can take on. We define that x must successively pass through the points of a certain "point set". The language used is therefore geometric, but thanks to the Cantor-Dedekind Axiom (p. [7]) the arithmetical meaning is defined as well.

Within this point set, y is called a function of x, in symbols $y = f(x)$, *if for every x of the set there is a specific y* (x and y are intended as precisely defined numbers, that is, as decimal fractions with well defined digits).

Usually, x passes continuously through the points of a part of the axis, which correspond to the set of all points between two fixed ones m and n. Such a point set is also called an *interval* $\overline{mn}$; we speak of a *closed interval* if the endpoints belong to the interval, of an *open interval* if the endpoints do not belong to the interval. We thus obtain the *older definition of a function*, like the one used for instance by *Lejeune-Dirichlet: y is called a function of x in an interval, if for every numerical value x in the interval, there is a well-defined numerical value of y.*

This should be the more general concept of function within precision mathematics. Once more I refer to the report of *Pringsheim* in the mathematical encyclopaedia:

Alfred Pringsheim: Allgemeine Funktionentheorie. *Enzyklopädie der mathematischen Wissenschaften*, Vol. 2 (finished in 1899), p. 1 to 53,

where there is a beautiful account of how the generalized concept of function has become accepted.

In my paper on the general concept of function, which I mentioned at the end of the introduction, I have already outlined what I am going to say now about the *concept of function in empirical mathematics*. You can see the following as a development of the ideas outlined there.

F. Klein, *Elementary Mathematics from a Higher Standpoint*,
DOI 10.1007/978-3-662-49439-4_3

Similar thoughts have been developed by *Moritz Pasch* 1882 in two papers that contain very interesting observations despite their brevity. These are the "Vorlesungen über neuere Geometrie", Leipzig 1882 (where Pasch starts with empirically given points and lines, and then moves up to abstract formulations[27]), and in particular the "Einleitung in die Differential- und Integralrechnung", Leipzig 1882. I
[15] particularly wish to bring this book to your attention ; you can also consider *Pasch's* parallel remarks in vol. 30 of the *Mathematische Annalen* of 1887, p. 127–131[28].

In recent times the question of the treatment of empirical geometry – in the sense of a science that can be controlled by exact experiments and that does not contradict reality – has been developed by the Danish mathematician Johannes Hjelmslev in different works:

1. Die Geometrie der Wirklichkeit. *Acta mathematica*, vol. 40 (1916), pp. 35–66;

2. Die natürliche Geometrie. *Abhandlungen aus dem mathematischen Seminar der Hamburgischen Universität*, vol. 2 (1923), pp. 1–36;

3. Elementaer Geometri (3 Voll.), Kopenhagen 1916–1921; a textbook that has been introduced in various schools of Denmark.

After this digression on literature, we address the question:

How might we be confronted with an equation $y = f(x)$ in empirical geometry?

Let us fix an orthogonal coordinate system and ask to what extent y – as a function of x – is given by a curve that we draw with a continuous stroke of the pen or that we imagine given by a chart recorder (which, for example, measures the temperature as a function of time) (Fig. 1). By the way, we can also fix a curve while not drawing with a continuous stroke but rather by stringing close together discontinuous single points one after the other (Fig. 2).

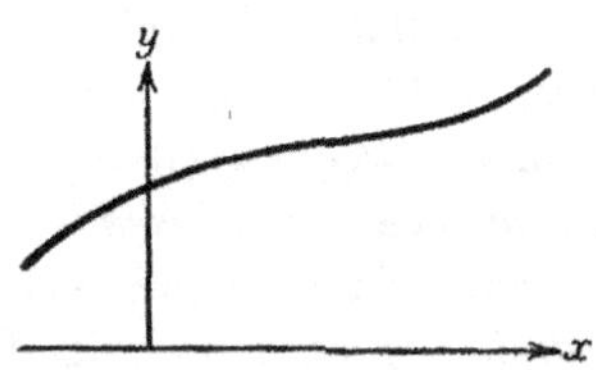

Figure 1

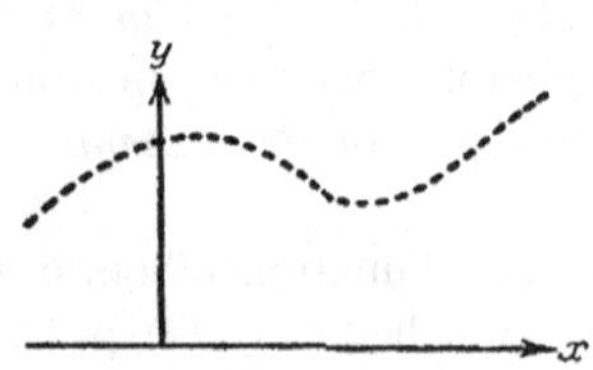

Figure 2

[27] [2nd edition with the appendix of *Max Dehn* about "Die Grundlegung der Geometrie in historischer Darstellung". Berlin: Julius Springer 1926.]

[28] Indeed, *Paul du Bois-Reymond* constantly considers in his "Allgemeine Funktionentheorie", Tübingen 1882, the difference between the "idealistic" and "empirical" conception (which he transfers to space intuition), but he does not turn this towards mathematics, instead he loses himself in spinning out the antinomies which he thinks arise from this distinction.

Surely, this collection of points appears, from close up, as discontinuous; nevertheless we have the impression that the points together identify a well-determined curve, and, from far away, we even believe to see a continuous curve.

What we can notice from this simple example is a *peculiarity of our visual sense*, which shows its effects also in other cases. Look, for instance, at a distant forest from which no tree in particular emerges; you will think you see a connected contour. Similarly you assume this characteristic of our vision when speaking of the **[16]**
contour of a human head with hair. So we must say: *When we look at a very large number of small objects that are close together the idea of continuum arises within us.*

Before, I spoke of a continuous stroke of the pen. However, should we look at it with a sufficiently sharp microscope it would be made up of many small stains. So we have to add to our last proposition: *Even if we believe to perceive a continuum in front of us, a more accurate observation will often convince us that we are only observing a dense succession of small particles.*

At first, I spoke of this peculiarity only in the case of the visual perception of close objects lying side by side. But the peculiarity also holds in the case of perception of time-successions. The fact that visual perceptions following each other in small intervals seem to merge into a continuum can be observed particularly well by the functioning of the cinematograph.

To investigate how this can be explained is a task for physiologists and psychologists. Here I only want to mention that the elements of our retina are composed of discontinuous adjacent little nerve rods, and probably each of them can reproduce only a certain partial perception, so that for us there is no difference between the observation of an effective continuum and the observation of a series of objects which follow each other in sufficiently small intervals. But I must obviously leave to the expert a real judgment of the pertinence of physiological and psychological studies to penetrate the facts of sensorial perception, investigating how the vision process is realised, the structure of the retina, the chemical reactions that light produces on the retina by irradiation, the duration of such reactions, etc. Here we refer only to the facts as such.

Following our considerations we now ask ourselves: how can we determine y as a function of x by means of an empirical curve?

We answer: x and y are both exactly determined only up to a threshold. Therefore: *The empirical curve does not determine y as a precise function of x, instead it defines the y corresponding to an x only with a limited precision.*

We can write this as

$$y = f(x) \pm \varepsilon,$$

where the undetermined ε (that we can equally consider as a function of x) satisfies the inequality $|\varepsilon| < \delta$, this means that the absolute value of the small quantity ε is **[17]**
below the threshold δ.

In the paper from 1873 that I mentioned, I called an object such as $y = f(x) \pm \varepsilon$, which has a certain width, a "function stripe", so now we can say briefly: A curve given empirically does not define a function, but a function stripe.

On the Efficiency of Space Intuition

As the shape of a function can be provided only with limited precision by an empirical curve, we advance a *psychological* question, which I already defined as controversial: Is it possible to exactly conceive a function by means of a spatially perceived curve?

This "imagining" of a curve has not to be mistaken with the understanding of the law that exactly defines the curve on the basis of absolutely correctly assumed axioms. It is rather the vivid perception of a concrete form.

My answer to the question is not surprising considering what I said before about the imaginability of sets of points on the x-axis; I already expressed this as follows in the mentioned note of 1873: *Even if a curve is not drawn nor it is assumed to be tracked by the eye, but we have a "perception" of it, it has in any case a limited precision and does not therefore correspond to the exact concept of a function of precision mathematics but rather to the idea of a function stripe.*

The strength of this view will become furthermore evident when I will exactly define, by means of the axioms, a great variety of curves that certainly cannot be imagined (with the same certainty that the law generating them can clearly be understood).

An opposite point of view is presented in mathematical literature, in particular by *Alfred Köpcke*[29] in the *Mathematische Annalen* vol. 29 (1887), pp. 123–140. I wish to speak about this, the more as I suppose that his point of view is widespread, even if not formulated as pointedly.

To understand *Köpcke*'s point of view, I need to anticipate something that I will treat later on in a more extensive way.

Among the functions, there exists a class of functions, the *continuous* functions, which includes the *differentiable* functions as a particular subclass. If differentiation is possible as often as can be desired, functions are called *indefinitely differentiable.*

Köpcke maintains that we can exactly imagine those curves $y = f(x)$, *for which* $f(x)$ *is not only continuous, but also indefinitely differentiable, while, on the other hand, we absolutely cannot imagine curves which correspond to continuous functions that do not have a derivative.*[30]

[18] You can see how this differs from my point of view. While I claim that spatial perception is always imprecise, Köpcke says that it is completely sufficient for a certain class of functions, but that it fails in the case of other kinds of functions.

One would think that physiologists and psychologists have engaged on this question. But this seems not to be the case; moreover, the only researcher who faced this question from the point of view of mathematics – besides Köpcke and myself – seems to be Pasch in his already mentioned books and in his paper in *Mathematische Annalen*, vol. 30 (1887). In fact, a very poor treatment is given to the most impor-

[29] Translator's note: There is a typing error in the German version: The name of the mathematician *Köpcke* is written as 'Köpke'.

[30] Translators' note: Klein uses the term "Differentialquotient", as is typical in German.

tant questions that pertain to the border areas of various sciences. In the present case, we have to also consider – with respect to the neglect of the question – that little is known outside the mathematical circles about the modern *arithmetisation* of mathematics (that is its logical construction on the basis of the modern concept of number) and even less is its importance understood.

On the Exactness of the Laws of Nature (with a Digression About Differing Conceptions Regarding the Constitution of Matter)

The more readily I like to devote myself to a short digression about *natural philosophy*, by posing the following question, which arises from the present context: *If we are not able to establish an exact function* $y = f(x)$ *empirically nor to perceive it ideally, what about the exactness of the so-called natural laws? Are they exact or approximate?*

Take any law, such as those treated in physics or in natural sciences, for instance the *law of falling bodies*

$$y = \frac{gt^2}{2} + ct + c'.$$

To what extent is this law confirmed by observation? Here too we must say: It is confirmed by observation only in an approximate way. Similarly, a thrown object does not describe an exact parabola; in fact the ballistic curve can sensibly differ from a parabola. And a tennis ball can describe a curve with an upward peak.

How do we come to formulate this law in an absolutely exact way while attributing all deviations to disturbing factors or observation errors?

Here we need to consider a new principle, the *principle of simplicity of nature's explanation* or of the *economy of thought*, which in the present situation can be expressed as follows:

For establishing the laws of nature one resorts (not deliberately but involuntarily) to the simplest formulas that seem to describe the phenomena with reasonable accuracy.

What we said here especially regarding the law of falling bodies is still valid for the natural laws that are apparently the best ascertained ones, for instance for the *laws of the constancy of mass and energy*.

In most experiments (particularly in the chemical ones) the mass is fairly constant; this is expressed by a formula [19]

$$M = \text{const} + \varepsilon(t),$$

where $\varepsilon(t)$, is a function of time which assumes only very small values, and of which nothing definite has been determined.

Something similar holds for energy. In experiments energy, too, is fairly constant, but there is no mention of absolute precision[31], so that we must say: *Even those laws of nature that are the most general and important for the world view have always been proved experimentally only in a confined ambit and with limited accuracy.* We would like to say:

The exact formulation of the laws of nature by simple formulas is based on the desire to master external phenomena with the simplest tools possible.

To make sure that a natural law does not become a dogma, in other words a theorem that we consider irrefutable without asking for further evidence, we should always pose the question from a scientific point of view: Within what limits of precision is the formula proved by observation? We will mostly discover that this happens with much less precision than is commonly believed.

I wish to explain what I said with an *example* taken from astronomy, which is among those empirical sciences, in which precision is pushed to the maximum. I refer to the following work:

Simon Newcomb: The elements of the four inner planets and the fundamental constants of Astronomy. Washington 1895.

In this book, *Newcomb* – thanks to his life's work on the motion of the great planets – gives the best values that can be obtained in today's astronomy for all the "astronomical constants", and at the same time he gives the margins within which he considers them as firmly determined[32].

On pp. 118 to 120 he reflects about *how exactly the Newtonian Law of mutual attraction of two masses*

$$f = \frac{kmm'}{r^2}, \quad (k \text{ is a fixed constant})$$

is confirmed by observation.

[20] He distinguishes three kinds of observations:

1. Observations on Earth, where r varies between a few centimetres (in laboratory experiments) and the length of the Earth's radius.

2. Observations in which the interval of r varies between the length of the Earth's radius and the distance from the Earth to the Sun (law of falling bodies, motions of the Moon and the Earth with respect to the Sun);

3. Observations, in which r increases up to twenty times the Earth-Sun distance (motion of the most distant planets).

With respect to the validity of the Newtonian law of gravitation, or rather to the precision with which this law is confirmed without doubt by observation in the three cases, *Newcomb* fixes the *uncertainty* as follows

---

[31] This precision is actually quite low and in any case much less significant than in the experiments by which the constancy of mass has to be proved.

[32] [A collection of recent data can be found in the paper of the Encyclopaedia (volume VI_2, 17; finished 1919) of *Julius Bauschinger*: Bestimmung und Zusammenhang der astronomischen Konstanten, where the constants are presented in a consequent manner as intervals, too.]

1. for the first mentioned interval, it equals $\frac{1}{3}$ of the value found for the force of attraction f,

2. for the interval between the length of the Earth's radius and the Earth-Sun distance it is $\frac{1}{5000}$ of the value found for f,

3. only for the intervals that are greater than the distance of the Earth from the most distant planets it is very small.

Obviously this uncertainty should not be interpreted in the sense that this law gives, for instance in the first interval, values which are wrong up to $\frac{1}{3}f$, but only in the sense that such an error seems possible due to the available experimental data.

As a curiosity I add here that, to take account of certain irregularities in the motion of Mercury[33], the American astronomer *Asaph Hall* states the Newtonian law as follows:

$$f = \frac{k m m'}{r^{2+0,1574 \cdot 10^{-7}}}.$$

After all, we have to say (and so my assertion is justified):

The precision with which the generally accepted laws of nature are proved to be compelling by means of an experiment is very limited even in the case of Newton's law of attraction.

Following these remarks, let me take up the natural-philosophical question addressed by the Faculty of Philosophy of Göttingen University in its report for the *Beneke prize* mentioned in the introduction.

Here, too, we essentially find considerations on the question of the difference between the approximate evaluation of magnitudes and the rigorous definition of number. But in particular the *question concerning the internal structure of the matter* is addressed (you will soon see that this has much to do with our subject). **[21]**

As for the question of the internal nature of matter there are two *extreme points of view*:

1. The first one considers that matter effectively fills space up in a continuous way (*theory of continuity*),

2. The second considers it as an accumulation of separate, strictly point-like masses (*extreme atomic theory*).

The opinions of most scientists today lie in the middle. They see the matter as constituted by molecules; the molecules are however not intended as points, but rather as single complicated worlds.

The first extreme point of view was once represented in particular by *Wilhelm*[34] *Ostwald* and his school (*energetics*), who rejected any atomic theory. The second conception is found mainly among the French mathematicians of the classical school, particularly in *Pierre S. Laplace*, and remained long dominant; for instance, it was still held to be a valid aim in the first publication of *Herrman von Helmholtz*

[33] [As is known, Einstein's theory of relativity has brought great progress and changes to this field. An extended critique of the *Newton*ian law of gravitation can be found in the paper of the Enzyklopädie (VI $_2$, 22 by *Samuel Oppenheim* (finished in 1920).]

[34] Translator's note: There is a typing error in the German edition: It was printed *F. Ostwald* instead of *W. Ostwald*.

in 1847. The assumption is that the separate "point-like masses" interact with "central forces". This corresponds to a generalization of the conception that was dominant in astronomy ever since Newton.

Incidentally, this second extreme point of view has been significantly enhanced. The starting question was: Since matter fills space only in a discontinuous way, why should not *time be discontinuous also*? So, must not also any movement – that takes place only in discontinuous time intervals – be considered as occurring in leaps? One is led to this question when considering the effect of the cinematograph, which dissembles us discontinuous processes as continuous ones. In fact, such opinions were supported by *William K. Clifford* (On theories of the physical forces, *Proceedings of the Royal Institution* 1870, printed in "Lectures and Essays", vol. I, pp. 120–138, London 1901) and *Ludwig Boltzmann*[35].

How can such diverse views coexist? This is the legitimate question, which evokes the genuine interest of the mathematician on these themes. We must state:

The different premises will possibly allow equally good explanations with respect to the imprecise nature of our sensory perception, because the sensory perception is, in fact, not concerned with issues of precision mathematics but of approximation mathematics.

The proposed formulation will be willingly accepted. However it basically anticipates the result of an earnest mathematical reflection.

[22] The mathematician has neither the ability nor the task to decide which is "the case of nature"; he must leave this to the competent scientist, assuming it is possible to reach any given result at all. But he *can and must* decide whether the different premises – based on suitable conditions – can, if applicable, lead to the same results. This is, in fact, a great task for approximation mathematics!

Properties of the Empirical Curve: Connectedness, Slope, Curvature

After these general considerations I turn back to the specific question:

Everyone connects certain properties with the idea of an empirical curve (drawn by hand, for instance). Is it possible to restrict the concept of exact curve, in our present context the function $y = f(x)$ of precision mathematics, so as to regain analogous properties?

I sequentially list here the properties I am thinking of:

1. A first property that we in fact ascribe to the empirical curve is *continuity*, that is, we have the perception that the curve is connected in its smallest parts.

This corresponds to the property that $f(x)$ passes through all the intermediate values, or better:

[35] [As is known, the mentioned idea of a discontinuous change of state has gained its full validity in the modern quantum theory.]

If $f(a) = A$ and $f(b) = B$, then $f(x)$ should actually take on all the values between A and B in the interval from $x = a$ to $x = b$. We will learn that the so called "continuous" functions have this property (Fig. 3).

2. The second property that we infer from the empirical behaviour of the curve is that it – linked to its continuity – encloses a specific *area* between the x-axis and the ordinates of two of its points (Fig. 4). Here we are thinking of curves that are cut in one point by a straight line, parallel to the y-axis.

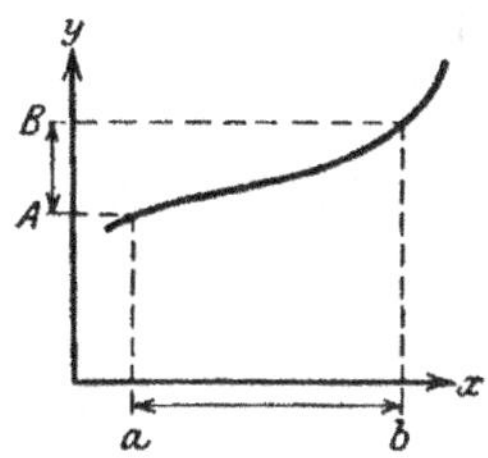

Figure 3

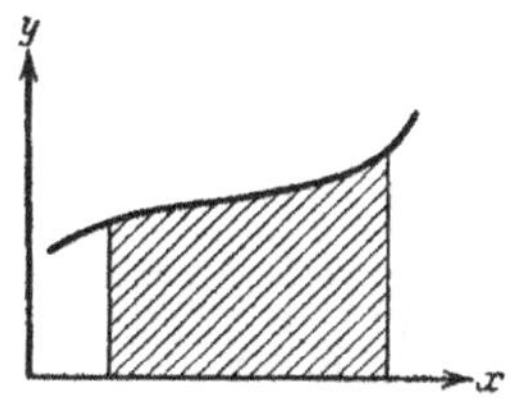

Figure 4

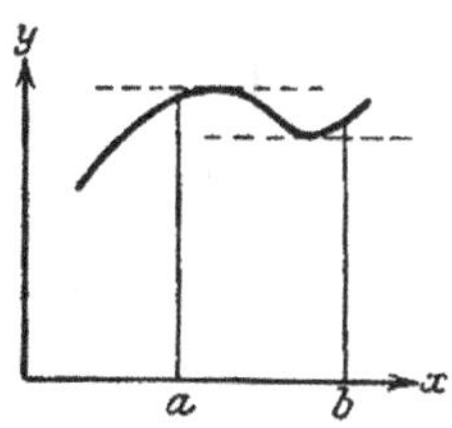

Figure 5

This does not correspond to a new restriction for $f(x)$. Rather we will find that the area, that is the definite integral $\int_a^b f(x)dx$, is well defined for every function that is continuous in the interval $a \leqq x \leqq b$, in other words, every continuous **[23]**
function is *integrable*.

3. Now we observe the general shape of the empirical curve, which is given by how the ordinates increase or decrease, and we find out that the ordinates assume only finite values, that the curve has its smallest or greatest value in the interior of the interval or at its endpoints, and that it presents only a finite number of maxima and minima in the interval (Fig. 5).

For an arithmetically defined continuous function $y = f(x)$ the maxima and minima can have an accumulation point inside a finite interval. If we consider, for instance, $y = \sin \frac{1}{x}$ (Fig. 6), the function oscillates in the interval from $x = \frac{2}{\pi}$ to $x = -\frac{2}{\pi}$ infinitely many times, between the values $y = +1$ and $y = -1$ (the number of oscillations increases when x is approaching zero).

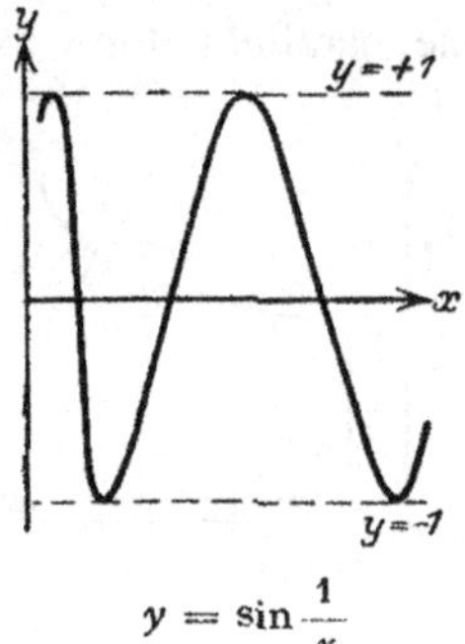

Figure 6

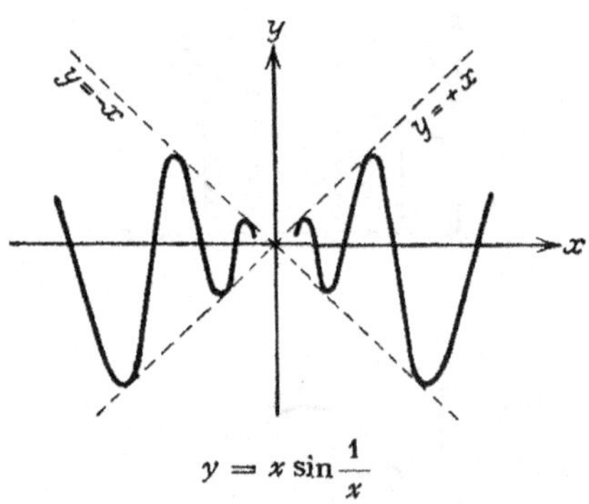

Figure 7

Another case is the function $y = x \sin \frac{1}{x}$, which has infinitely many maxima and minima in a finite interval that contains the origin (Fig. 7). This second example has the advantage that the function is not only continuous to the right and left of zero, but also at the origin.

Therefore, as it is not excluded that a continuous function has infinitely many maxima and minima in a finite interval, we have to make a particular request if we want to achieve conformity with an empirical curve:

The function $y = f(x)$ must split – in the given interval – into a finite number of monotonous parts.

4. We assign to an empirical curve everywhere a *slope*[36]. What do we mean by slope? We will try to give a preferably unbiased answer, including nothing in the definition that we already know perhaps from differential calculus.

---

[36] Translator's note: Klein uses the more general term *Richtung* (direction), which is not characteristic of calculus. Anyway, here the meaning is *slope*.

The drawn curve may represent a stream of finite width, so that the ordinate is determined only up to a threshold δ (if the width of the stream is significant, we **[24]** need to refer the following to a narrower stripe in the middle). Now, in order to determine the slope, the practical procedure is the following:

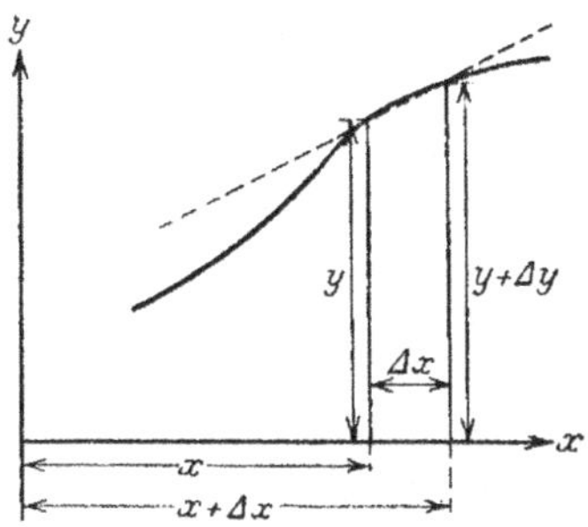

Figure 8

We take on the x-axis a small interval Δx (see Fig. 8), which is large with respect to the width of the stream, but small with respect to its entire course, we find for x and $x+\Delta x$ the corresponding ordinates y and $y+\Delta y$ and connect the two endpoints with a straight line. Of course, this line is not determined in an exact way, as the two points that it is connecting are not exact. But this straight line gives us what we call in practice the slope at the considered point, which can be determined only with a limited precision. We measure the slope[37] by means of the difference quotient $\frac{\Delta y}{\Delta x}$.

So we can say briefly:

An empirical curve has at every point a slope, which is determined by the difference quotient $\frac{\Delta y}{\Delta x}$ *depending on a certain threshold to be fixed in each particular case and where* $\Delta x \neq 0$ *is a quantity, which is small with respect to the curve's entire course, but large with respect to its threshold. The empirical curve approximately matches the line connecting the points* (x, y) and $(x + \Delta x, y + \Delta y)$.

We must now consider the question whether *every continuous function has a derivative.*

We define the derivative as

$$\lim_{\Delta x \to 0} \frac{\Delta y}{\Delta x}, \qquad \Delta x \neq 0$$

and we write briefly $\frac{dy}{dx}$ where d has to suggest that we consider the passage to the limit as achieved. But we have to consider that the derivative arises from the difference quotient discussed above by means of *two* limit passages which have to be performed in a definite order, by first reducing the width of the stripe indefinitely and then the interval Δx of the x-axis. As a general principle about passages to the limit I add here: *It is impossible to say a priori what the result of a limit passage*

[37] More in general, we could put $\frac{\Delta y}{\Delta x} = \frac{y_1 - y_2}{x_1 - x_2}$, where x_1 and x_2 are taken at an adequate distance at the left and at the right of x.

will be. But if there are several limit passages it is particularly important to observe their order exactly.

[25] Previously, mathematicians did not always follow this principle and they believed that the derivative of every continuous function existed. But from the studies of *Bernard Bolzano, Bernhard Riemann* and *Karl Weierstraß* we know that continuous functions as such do not need to always have a definite derivative, that is, it is possible that for $\frac{\Delta y}{\Delta x}$, such a limit as we just defined it, does not exist. We will explain this later on in more detail.

How did the contrary view arise? We do not know for sure where the fallacy resided. But it seems that the amiss assumption, that a continuous function should always have a definite derivative, arose from the fact that in our empirical curve the threshold and the small interval Δx were conceived as decreasing *simultaneously*, in such a way that the essential features of the figure remained, that is in the interval Δx the curve was assumed to maintain the conformity with a straight line. So, also for a rigorous function, the possibility to approximate a small part of it by a straight line was maintained.

If this interpretation is correct, *the fallacy lies in the fact that the two passages to the limit were performed at the same time and at the same rate, while it is required for them to be performed one after the other.*

Later we will explain in more detail the possibility of the non-existence of the derivative for continuous functions. Now it suffices to know that a continuous function does not need to have a derivative and, therefore, we need *to require the existence of a derivative for continuous functions as a premise* in order to maintain the conformity with our empirical curve, which clearly must have a slope.

5. It is also obvious that every empirical curve has a *curvature.*

The curvature of a curve is defined as the reciprocal $\frac{1}{\varrho}$ of the radius ϱ of a circle that passes through three points of the curve chosen appropriately. If the three points are[38]

$$x, y; \quad x + \Delta x, y + \Delta y; \quad x + 2\Delta x, y + 2\Delta y + \Delta^2 y,$$

for the curvature we have to consider the second difference $\Delta^2 y$ or – if we divide by $(\Delta x)^2$ – the second difference quotient $\frac{\Delta^2 y}{(\Delta x)^2}$. This gives the measure of the deviation from the straight line ($\Delta^2 y = 0$); we can also write it in the form

$$\frac{f(x + 2\Delta x) - 2f(x + \Delta x) + f(x)}{(\Delta x)^2}$$

How can we choose the three points in practice? Obviously, again we must stay within intervals, which are large with respect to the width of the stripe and
[26] small with respect to its total length. The precision with which we can define the second difference quotient – and consequently the curvature – of a stripe given empirically in accordance with our settings is smaller than the precision with which

[38] Translator's note: If we put: $\Delta y^* = f(x + 2\Delta x) - f(x + \Delta x)$, then $\Delta^2 y = \Delta y^* - \Delta y$. In this case Klein seems not to consider necessary to explain the meaning of the second difference quotient.

we determined the first difference quotient, and therefore the slope of the curve; indeed, the three points can now be moved on the stripe.

This indeterminacy increases if we go on to the third, fourth, ... difference quotient. For precision mathematics, in order to describe the curve we need to *postulate* the existence of the second and of higher order derivatives in order to maintain the analogy with the empirical curve.

I summarize points 4. and 5. as follows:

In order to regain in a rigorously defined function those properties that are analogous to those ascribed to an empirical curve with respect to slope and curvature (first and higher difference quotients), we need not only to require that the function is continuous and has a finite number of maxima and minima in a finite interval, but also assume explicitly that it has the first and a series of higher derivatives (as many as one will want to use)[39].

Cauchy's definition of the Continuous Curve. How Far-Reaching is the Analogy with an Empirical Curve?

For what will follow now, our course program is determined by these last developments. We need to define the properties, ascribed to the curves as gathered from experience, for the functions $y = f(x)$ in a rigorous manner, from the point of view of precision mathematics.

1. First of all the *definition of continuity*.

I specify that historically this definition, as most of the definitions that we are going to consider, made its way into science through the following fundamental works:

August Louis Cauchy: Cours d'analyse Vol. 1. Paris 1821, and *Résume des Leçons données sur le calcul infinitésimal*, Paris 1823,
in which *Cauchy* (1789–1857) elaborated his lessons given at the *École polytechnique*.

Independently of *Cauchy*, another researcher was active in this field, whose name **[27]**
became widely known only later. It is *Bolzano* (1781–1848), in Prague from 1817. To learn about his role in the history of differential and integral calculus you can

[39] Of course we could make other requirements for the conformance. In an empirical curve the slope, the curvature, etc., pre-determine a part of the further development (what is in so far evident, as it is through this development that we determine the slope, the curvature, ... in a point). *Alfred Köpcke* from Ottensen (whom I mentioned on p. 17 and with whom I spoke some time extensively at the *Naturforscherversammlung*) considers it appropriate to hold this *pre-determination* also in functions $y = f(x)$ in precision mathematics. So he obviously arrives at classes of functions that have more restrictions than those considered in this text. For such selected functions *Köpcke* claims there is a greater degree of intuitiveness than for the others. Of course I object since in my opinion no function is intuitive at all, but only function stripes. (Translator's note: Köpcke was printed here erroneously as 'Otto Köpke' in the 1928 edition. The *Naturforscherversammlung* (meeting of the natural scientists) was in 1901).

see the work of *Otto Stolz* in *Mathematische Annalen* vol. 18 (1881), pp. 255–279: Über Bolzanos Bedeutung für die Geschichte der Infinitesimalrechnung.

As often happens in science, here too, an historical development is tied only to one name, while such a development was possibly arising simultaneously in various locations.

Now, what is *Cauchy's definition of continuity*?

A function $y = f(x)$ is said to be continuous at a point $x = x_0$ firstly if it is defined unequivocally at this point, and secondly if for every positive number η, however small, you can find a positive number ξ such that for all points x, x' internal to any interval which contains the point x_0, you get $|f(x) - f(x')| < \eta$ provided that $|x - x'| < \xi$. Expressed in a more intuitive way: However you choose η, it must always be possible to find in a neighbourhood of x_0 an interval whose length $\delta > 0$ is such that for all pairs x, x' inside this interval, the difference in absolute value of the respective function values is smaller than η.

If the condition $|f(x) - f(x')| < \eta$ is met only by intervals for which x_0 is an endpoint, we say that the function $y = f(x)$ is one-sidedly continuous; in particular we call a function right-continuous or left-continuous depending on the position of x_0[40]. Now we can easily define what is intended by "a function continuous in an interval". We mean a function that is continuous at every point of the interval; in the case of a closed interval, however, we will require left-continuity for the right endpoint and right-continuity for the left endpoint.

Thanks to this definition, we will prove *that a function that is continuous in a closed interval $a \leqq x \leqq b$ takes on all values between $f(a)$ and $f(b)$.*[41]

We prove the theorem in a special case, which however sheds light on the general principle. Let us consider $f(x)$ at two points with integer abscissas $x = a$ and $x = a + 1$, where $f(a) > 0$ and $f(a + 1) < 0$ (Fig. 9). We prove that in this case $f(x)$ equals zero at least once in the interval a and $a + 1$.

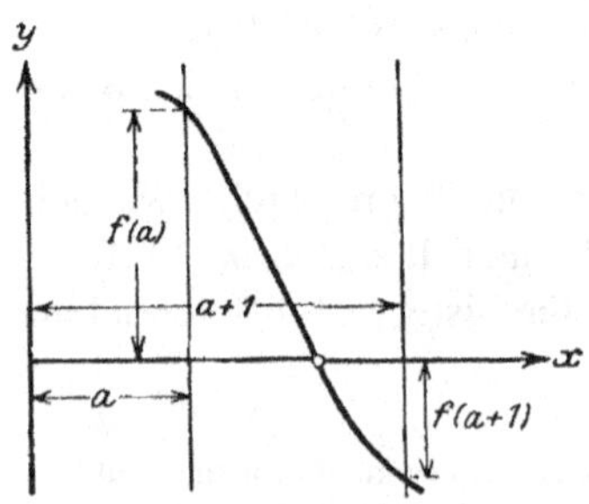

Figure 9

[40] Translators' note: We use here the usual English terminology; Klein, however, applied a geometrically loaded terminology: "vorwärts stetig" – ahead continuous – and "rückwärts stetig" – reversely continuous.

[41] Translator's note: Only in the subject index Klein mentions the *intermediate value theorem* with reference to pp. 27–29.

Intuitively it is clear that a continuous curve which is in a above the x-axis and in $a+1$ below the x-axis crosses the axis at least once. *But theoretically we must prove that the theorem considered can be derived from the given definition of continuity, with reference to the number concept fixed once and for all.* [28]

Expressed differently: We want to verify the effectiveness of Cauchy's definition by proving that it implies just that theorem, which immediately follows from the continuity of the empirical curve (because of our natural ideas on continuity).

So, the theorem that we want to prove is, again: *A function, which is continuous in the sense of Cauchy in the interval $a \leqq x \leqq a+1$, that is positive at $x = a$ and negative at $x = a+1$, has at least one zero between a and $a+1$.*

The proof has the same format as our proofs about sets of points, it is a "proof by decimals".

We divide the interval between $a = a_0$ and $a_0 + 1$ in 10 equal parts by one-digit decimals

$$a_0, 0;\;\; a_0, 1;\;\; a_0, 2; \ldots;\;\; a_0, 9;\;\; a_0 + 1{,}0.$$

If it is not the case that for one of these decimals $f(x)$ is equal to zero (we exclude this case, because otherwise we would have nothing to prove), then there exists a first interval $a_0, a_1 \ldots a_0, a_1 + 1$, for which at its left endpoint $f(x)$ is still positive, and for which at its right endpoint $f(x)$ is negative, so that exactly the same holds in this tenth as it does in the interval with integer endpoints.

We now divide this tenth of the starting interval $a_0, a_1 \ldots a_0, a_1 + 1$ into 10 hundredths of the starting interval

$$a_0, a_1 0;\;\; a_0, a_1 1;\;\; a_0, a_1 2; \ldots;\;\; a_0, a_1 9;\;\; a_0, a_1 + 0,$$

where again $f(x)$ can became zero in one of the dividing points, which we want to exclude. We choose again the first of these intervals for which at the left endpoint $f(x)$ is positive and at the right endpoint $f(x)$ is negative. As we can easily understand, from the ceaseless repetition of this process we obtain nested intervals, that is, an infinite sequence of intervals each of which is contained in all the preceding ones and whose length tends to zero. The two sequences of decimal fractions constituted by the left and the right endpoints of the intervals have therefore the same decimal fraction x_0 as their limit. We want to prove that, according to Cauchy's definition of continuity, $f(x_0)$ becomes equal to zero. This theorem can be reduced to a *general principle*: x_0 is the common limit of a decreasing and of an increasing sequence of decimal fractions. For the first one $f(x)$ is positive, for the second one $f(x)$ is negative

$$\left.\begin{array}{l} x_0 = \lim a_0 | a_0, a_1 | a_0, a_1 a_2 | \ldots \text{ with } f(x) > 0, \\ x_0 = \lim a_0 + 1 | a_0, a_1 + 1 | a_0, a_1 a_2 + 1 | \ldots \text{ with } f(x) < 0, \end{array}\right\} \begin{array}{l}\text{for all } x\text{, that} \\ \text{belong to the} \\ \text{sequence.}\end{array}$$

For continuous functions we have the theorem: [29]

If x_0 is the limit of a sequence of points $x_1, x_2, \ldots$, then also $f(x_0)$ is the limit of the function values $f(x_1), f(x_2) \ldots$

This follows easily from the content of Cauchy's definition of continuity and from the definition of limit.

This immediately proves our previous theorem: Because $f(x_0)$ appears once as the limit of a sequence of only positive numbers, and again as the limit of a sequence of only negative numbers. The two facts are compatible only if $f(x_0) = 0$.

At the end of the proof we used a proposition that is nearly more important than the theorem proved.

In symbols it is:

If $f(x)$ is continuous, then

$$f(\lim x) = \lim f(x).$$

From this, we can derive a further theorem that will be useful later on.

We start choosing as our point set the set of all points $x = \frac{m}{n}$ with rational abscissas. From that we can determine all irrational numbers, namely as limits of infinite sequences of rational numbers. So, for an irrational x_0 we have

$$f(x_0) = f(\lim \text{rat. } x) = \lim f(\text{rat. } x),$$

in words: *The course of the values of a continuous function is determined at all points of an interval, if only it is determined for all rational points of this interval.*

We can immediately generalize this result, if we have in mind the point set – introduced earlier – that is *everywhere dense* in an interval (a, b). Such a set exists if each subinterval of (a, b), however small, contains points of the set. We see that for such a set all the points in (a, b) are accumulation points and therefore for every point x_0 of the interval (a, b) – even if it does not belong to the set – one can find sequences of points of the set which have x_0 as their limit. So we obtain the theorem: *A continuous function is defined at all points of an interval if it is defined at all points of a set which is everywhere dense in the interval.*

For example, we can also choose the set of all finite decimal fractions (being sufficient to determine on the x-axis an everywhere dense point set) so that we obtain as a corollary:

A continuous function is completely defined if it is defined for all the numbers x that can be written as finite decimal fractions.

2. Now we arrive at the second point, in which we speak of the *boundary of a surface*.

We have the impression, that every part of an empirical curve encircles an area
[30] bounded by the x-axis and the ordinates of its endpoints. In order to approximately determine it, we can cut out the paper surface, weigh it and compare it with the weight of a known area, for instance $100\,\text{cm}^2$. With this method we can often achieve reasonable results.

A further practical method is *mechanical quadrature or numerical integration*, to which also the *trapezium rule* and the *Simpson rule* belong.

To refer only to the trapezium rule, we divide the interval into intervals Δx (which are conveniently chosen of equal length) that are so small that the parts of curve between the endpoints of the ordinates of the single Δx can be considered approximately straight and the corresponding surface can consequently be calculated as the area of a trapezium (Fig. 10). By summing up all the areas of the trapezium we find that the area encircled by the curve is:

$$\Delta x \frac{y_0 + y_1}{2} + \Delta x \frac{y_1 + y_2}{2} + \cdots + \Delta x \frac{y_{n-1} + y_n}{2}$$
$$= \frac{\Delta x}{2} (y_0 + 2y_1 + \cdots + 2y_{n-1} + y_n) .$$

A third practical method is based on the use of a *planimeter*[42].

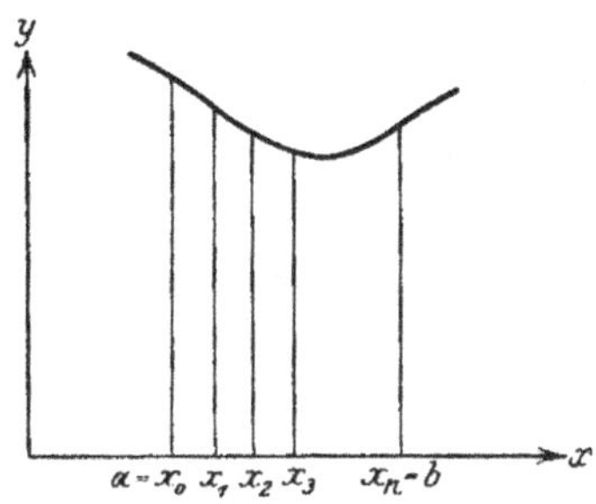

Figure 10

In the practical determination of an area, the major errors are not due to the substitution of the surface with a sum of rectangles or to the arbitrariness of the choice of Δx, but mostly to the width of the drawn or otherwise empirically given curve that delimits the surface. So, for instance, with a length of 10 cm and a width of $\frac{1}{3}$ mm of the stripe, the error that originates from the width is $\frac{1}{3}$ cm^2.

The Integrability of Continuous Functions

How do we interpret these findings *theoretically*?

The theoretical observation begins with the sum $S = \sum_1^n \Delta x_\mu \cdot f_\mu(x)$ where n is the number of intervals in any subdivision of the interval $a \leqq x \leqq b$ and $f_\mu(x)$ represents any ordinate in the subinterval Δx_μ, and we continue to observe the behaviour of this sum when the maximal length Δ_n of the subintervals tends to zero and hence evidently n increases indefinitely. If for *each* sequence of subdivisions that satisfies the condition $\Delta_n \to 0$ and for each choice of $f_\mu(x)$ there exists a limit, then we denote it with $\int_a^b f(x)dx$ and call it the *definite integral of $f(x)$ between the endpoints a and b.* The problem that we have to solve hereafter is:

[42] See, for instance, Vol. II, pp. [11]–[16].

[31] *When does such a limit exist? That is, when is the function $f(x)$ integrable?*

The first to pose this question and to face it in a general form was *Bernhard Riemann* in his *Habilitation* treatise (1854): Über die Darstellbarkeit einer willkürlichen Funktion durch eine trigonometrische Reihe[43].

We start by considering a function that is *bounded* within an interval $a \leqq x \leqq b$, that is a function whose function values do not exceed a given positive number *K*. This function does not need to be continuous. As we will prove later, in the whole interval as well as in any subinterval, the function has an upper and a lower limit. If G_μ is the upper limit and g_μ the lower limit in the subinterval Δx_μ (Fig. 11), then $f_\mu(x)$ is between g_μ and G_μ. So we find that: *Our initial sum $\Sigma \Delta x_\mu f_\mu(x)$ is between the two limits $\Sigma \Delta x_\mu G_\mu$ and $\Sigma \Delta x_\mu g_\mu$* and these two sums differ by $\Sigma \Delta x_\mu D_\mu$ where $D_\mu = G_\mu - g_\mu$ is called the *variation* of the function $f(x)$ in the μ-th subinterval.

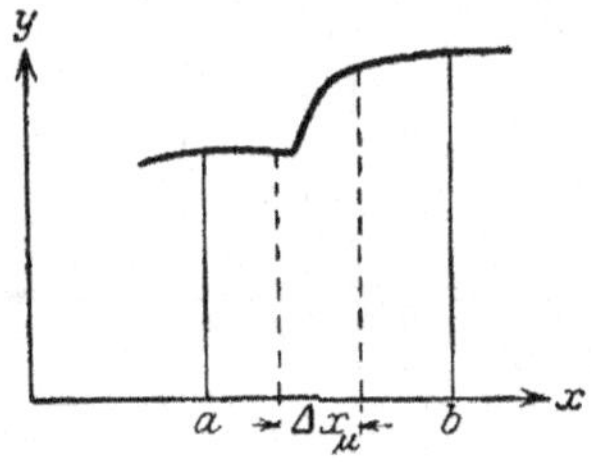

Figure 11

We can also prove – but we do not want to develop this point – that for bounded functions the two sums $\Sigma \Delta x_\mu G_\mu$ and $\Sigma \Delta x_\mu g_\mu$ tend for $\Delta_n \to 0$ to two limits *A* and *B*, which must also always include $\Sigma f_\mu(x_\mu) \Delta x_\mu$. If the two sums tend to the same limit *C*[44], the limit of $\Sigma f_\mu(x_\mu) \Delta x_\mu$ also exists and must be equal to *C*. But *A* and *B* are surely equal – and *the bounded function $f(x)$ is therefore integrable* – if $\Sigma \Delta x_\mu D_\mu$ can be made arbitrarily small by sufficiently reducing the maximal length Δ_μ of the subintervals.[45]

In particular we claim that: *For every function that is continuous in a closed interval $a \leqq x \leqq b$ – such a function is necessarily bounded – the sum $\Sigma \Delta x_\mu D_\mu$ can be rendered arbitrarily small.*

Before presenting the proof I need to mention *a subtle question concerning continuity*, which I shall not deepen anyway:

We called a function $f(x)$ continuous at the point x_0 if for every η there exists a ξ so that for all x with

$$|x - x_0| < \xi, \quad |f(x) - f(x_0)| < \eta$$

---

43 *Bernhard Riemann*, Gesammelte mathematische Werke und wissenschaftlicher Nachlass. Leipzig: Teubner, 1892, pp. 213–251

44 Translator's note: Actually, the text says literally: "If *A* and *B* have the same limit value *C*...".

45 Translator's note: This paragraph was printed with a smaller font and less space.

Now, how does continuity change if x_0 moves within the interval $a, \ldots b$, that is, *what is the measure of continuity* if x_0 passes across the interval?

We come across such a question if we consider a curve that increases slowly at a certain moment, and at another moment in a steep manner. At a point of relative flatness x_{01} I can obviously give for the same η a larger ξ then at a point of steepness x_{02} so that, once η is given, ξ can differ greatly for different points depending [32]
on whether the curve is flatter or steeper (Fig. 12).

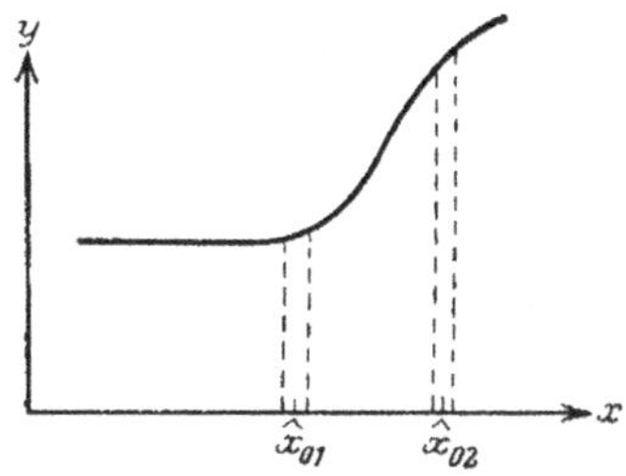

Figure 12

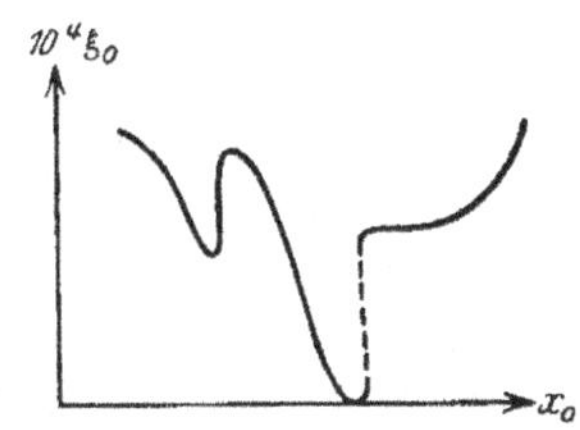

Figure 13

We can gain a clearer understanding of this by means of an *auxiliary curve* that allows us to evaluate the measure of continuity.[46]

Let us choose the abscissa x_0 and approximately $10^4\,\xi_0$ as the ordinate. The curve that originates (which reveals for a given η the dependence of ξ_0 on x_0) has positive ordinates over the whole interval. But it is not yet excluded that the shape of the auxiliary curve is the one suggested in Fig. 13. The ordinates could in fact come near to zero in one or more points, and suddenly jump up high at points where – due to continuity – they should become zero. In this way the requirement that the function must always have positive values would be satisfied. The auxiliary curve would then have a discontinuity. If we found such a behaviour, we would call the original function "not uniformly continuous" in the interval.

[46] Translator's note: The notion of uniform continuity was quite new in 1902. Klein wanted to give a proper introduction of non-uniform continuity, which nowadays tends to be given briefly only as negation of the uniform one. In introducing it, Klein exposes a thought experiment to motivate the notion; he follows the definition given by *Ulisse Dini* in 1878, in his "Fondamenti per la teorica delle funzioni di variabili reali", p. 47 (German translation in 1892). The thought experiment should show that non-uniform continuity can only occur near a point of discontinuity. See also Pringsheim's paper in the *Enzyklopädie* (see below, p. 33).

We shall say:

The assumption that a function is continuous at any single point of the interval seems not to exclude continuity being not uniform over the whole interval, that is, that there are points, at which ξ becomes smaller than any given quantity however small.

On the contrary, we have a *uniform continuity* in the interval if the ξ-curve – though fluctuating up and down – does not present such an exception (Fig. 14); that is, if for a fixed η we can give a ξ in such a way that for every pair of points x, x_0 of the interval as soon as $|x - x_0| < \xi$ we have also simultaneously that $|f(x) - f(x_0)| < \eta$.

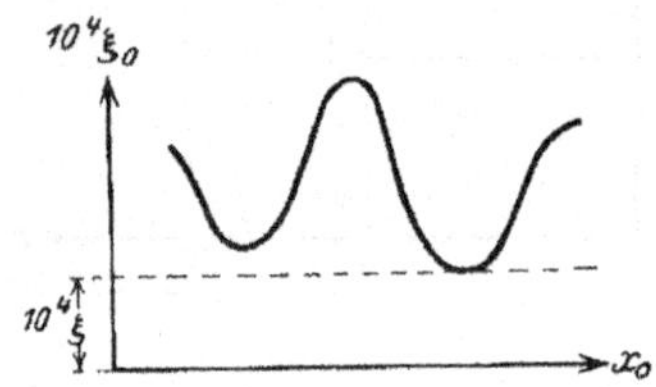

Figure 14

Now we arrive at the point whose proof I need to skip for the sake of brevity. We can show, that the unpleasant possibility of non-uniform continuity that we were incidentally thinking of is actually eliminated from the start for functions that are continuous in a closed interval, because a closed interval is *compact*, that is all its sequences of points have an accumulation point belonging to the interval.

So the following theorem holds: *Every function that is continuous in a closed interval is also uniformly continuous in it.*

To learn more about this I refer again to Alfred Pringsheim:

[33] Allgemeine Funktionentheorie. *Enzyklopädie der mathematischen Wissenschaften*, Vol. 2, p. 18[47]

I particularly insisted here on the possibility of non-uniform continuity because we will later deal with questions of convergence of series where we have to expect analogous possibilities.

Now I again take up the question of the integrability of continuous functions.

As we know that $f(x)$ is uniformly continuous in the closed interval $a \leqq x \leqq b$, we can find once and for all a value ξ for the *whole* interval for which, when $|x - x_0| < \xi$, we have $|f(x) - f(x_0)| < \eta$.

So, if we choose the length $\Delta x_\mu < \xi$, in the interval Δx_μ the variation becomes $G_\mu - g_\mu < \eta$ or $D_\mu < \eta$.

[47] [See also *Alfred Pringsheim*: Vorlesungen über Zahlen- und Funktionenlehre II, 1. Leipzig 1925, pp. 54–56 and *Felix Hausdorff*, Grundzüge der Mengenlehre, Second edition, Berlin 1927, Chap. 8, p. 197.]

For the sum of the variations multiplied by Δx_μ it now follows that:

$$\sum \Delta x_\mu D_\mu < \eta \sum \Delta x_\mu = \eta(b-a),$$

that means: If a function $f(x)$ is *continuous in a closed interval* we can take $\Sigma \Delta x_\mu D_\mu$ smaller than $\eta(b-a)$ and consequently smaller than any given positive number; *therefore such a function is always integrable.*

The Theorem of the Existence of the Greatest and of the Smallest Value

3. In this paragraph we consider the question of the *greatest and the smallest* values of a continuous function.

The ordinate of an empirical curve always has a greatest and a smallest value, which in certain circumstances has to be sought at the endpoints (Fig. 15) of an interval. The question is: Does an idealised curve, that is the function $f(x)$, which must obviously be bounded in the interval, have a greatest value in the interval $a \ldots b$ in the sense of precision mathematics?

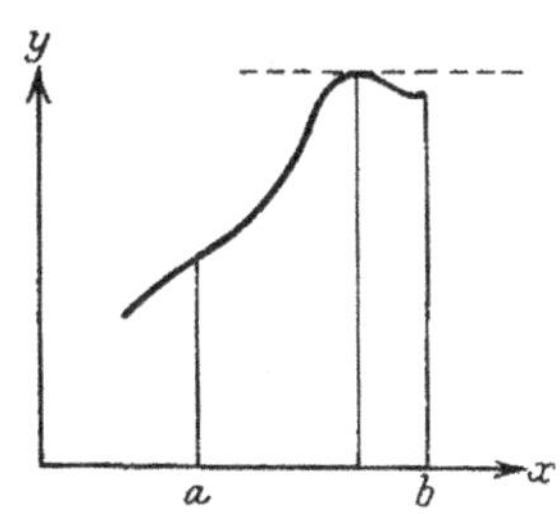

Figure 15

To begin with, we have here to distinguish between *upper limit and greatest value*. For the first one the theorem holds: *If the function $f(x)$ is bounded in the interval $a \leqq x \leqq b$ then the values of y in the interval surely have an upper (and a lower) limit*, as was already observed on p. 31.

The proof is a simple consequence of our first theorem about point sets, pp. [11]–[12]. In fact: For every x of our interval we have a particular y, such that we obtain on the y-axis a well-defined set of points that corresponds to the interval of the x-axis, and which is bounded from above.

For such a set of points we have proved the existence of an upper limit.

But for the moment it is not known if this upper limit belongs to the point set, that **[34]**
is, if it is or is not the ordinate of a given x, and so if the upper limit can be reached or not. I first want to give an example in which this is not the case. Suppose $f(x)$ is

defined for all points between $+1$ and -1 through the points of a semicircle whose centre lies in the origin and whose radius is 1; only for $x = 0$ let us have $f(x) = 0$:

$$y = \left|\sqrt{1-x^2}\right| \quad \text{for } x \neq 0,$$
$$y = 0 \quad \text{for } x = 0,$$

as indicated in Fig. 16. It is a well-defined function, which is nevertheless discontinuous. The upper limit is 1, but this value cannot be reached, because, according to the definition, at $x = 0$, where the value might be reached, we actually have $f(0) = 0$.

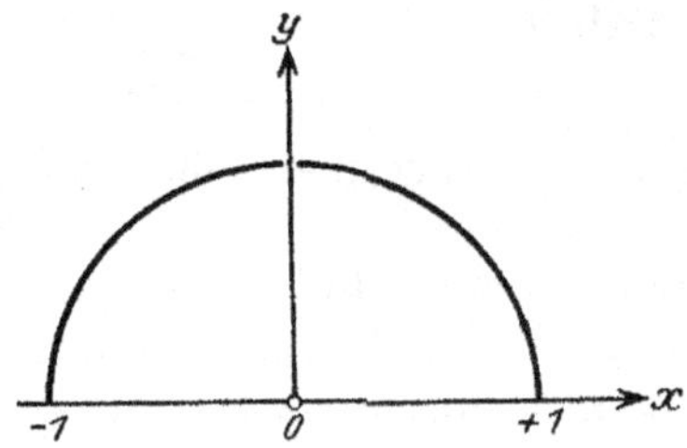

Figure 16

On the contrary, a function that is continuous in the interval $a \leqq x \leqq b$ reaches its upper limit, in other words *a function that is continuous in the interval* $a \leqq x \leqq b$ *has a greatest value*. For the validity of this theorem it is essential that the interval includes its endpoints. *Weierstraß* always particularly stressed this theorem in his lectures.

First of all: A function continuous in a closed interval is bounded from above, it therefore has an upper limit G. Now, with the help of the procedure by decimal interval nesting, we can infer the existence of a point x_0 with the property that in every neighbourhood of it, however small, G is the upper limit. We divide the initial interval in ten equal parts and choose among the subintervals a first one having G as its upper limit. We repeat the procedure with this subinterval. The ceaseless repetition of this procedure brings us to nested intervals defining a point x_0 of the desired type. So far, this can be proved for every function f that is bounded from above, *but* $f(x_0) = G$ *does not hold for every bounded function; this is true, however, without exception, for all functions that are continuous in a closed interval.* We now will expressly prove the latter.

We will do this indirectly, by supposing $f(x_0) = G'$, where G' is smaller than G. From the continuity of the function $f(x)$ we deduce that we can determine an interval containing x_0 so that the difference is

$$|f(x) - f(x_0)| \leqq \eta$$

where we chose η expressly smaller than $G - G'$. So, the values of the function lie
[35] between $G' - \eta$ and $G' + \eta$. But, as $G' + \eta$ is still smaller than G, it follows that G is

not an upper limit for our interval, which contradicts the definition of the point x_0. This contradiction can be formulated as follows: If we had $f(x_0) = G' < G$, we could – because of the continuity of $f(x)$ – determine an interval containing x_0 so that all the values of the function were smaller than G, at least for a fixed positive number δ, which is incompatible with the assumption that G is the upper limit for each interval containing the point x_0.

Summing up, the results are: Continuity that we assume as given for an empirical curve is found again in precision mathematics for those functions which are continuous in a closed interval and also for which the properties discussed hold, namely the encircling of a surface and the existence of a maximum and a minimum value.

As was already underlined, for conformance with empirical functions we will also explicitly assume that there exists, for our $f(x)$ only a finite number of maxima and minima, that is, of greatest and smallest values within each finite interval, with regard to a certain albeit maybe small neighbourhood.

The Four Derivates[48]

4. What about *the existence of a derivative* for our $f(x)$?

We recall, first of all: We considered the slope of the empirical curve to be defined by the difference quotient $\frac{\Delta y}{\Delta x}$ (where Δx has a specific order of magnitude). For the function $f(x)$ in precision mathematics we defined it by means of the derivative

$$\frac{dy}{dx} = \lim_{\Delta x \to 0} \frac{\Delta y}{\Delta x}.$$

The question is: *Does such a limit exist, and is it precisely the same limit for all the sequences Δx_υ converging to zero?*

To be completely clear I start with an *example*.

We choose the function we have already considered (Fig. 7, p. [23])

$$y = x \sin \frac{1}{x}.$$

It runs with always denser oscillations between the two straight lines $y = +x$ and $y = -x$ from the right and from the left towards zero, where it is, by the way, continuous (compare the wording of Cauchy's definition). Let us calculate the difference quotient for the origin and an arbitrary other point (x, y). It gives, of course

$$\frac{\Delta y}{\Delta x} = \frac{y}{x} = \sin \frac{1}{x}.$$

What can we say about the derivative?

---

[48] Translator's note: The type of derivative described in this chapter is often referred to as *derivate* or *Dini-derivative*, named after the mathematician *Ulisse Dini*. Klein uses the term "Derivierte".

[36] $\sin\frac{1}{x}$ is a function that oscillates infinitely often between $+1$ and -1 when x becomes indefinitely smaller.

Therefore: For Δx approaching zero, the difference quotient $\frac{\Delta y}{\Delta x}$ always oscillates between $+1$ and -1 without approaching a definite limit. *The numbers* $+1$ *and* -1 *represent the upper and the lower limits of the set of values taken on by the difference quotient, however small we might assume the interval in which* Δx *is varying to be.* We recognize that here the existence and the value of the limit of the sequence $\frac{\Delta y_v}{\Delta x_v}$ depends on the choice of the sequence Δx_v. If one considers only those sequences Δx_v for which the limit exists, it can take on any value from -1 up to $+1$, including the extreme values.

For any positive δ we denote the upper limit of all $\frac{\Delta y}{\Delta x}$ that we can form in the interval $0 < x < \delta$ for the point 0 by $G(0,\delta)$, the lower limit by $g(0,\delta)$, and the limit values $\lim\limits_{\delta\to 0} G\,(0,\delta)$, $\lim\limits_{\delta\to 0} g\,(0,\delta)$ by D^+ and D_+. Similarly the corresponding limits for negative δ are denoted by D^- and D_-. So in our example we have

$$D^+ = +1,\ \ D_+ = -1,\ \ D^- = +1,\ \ D_- = -1.$$

The numbers thus defined are called the four *derivates* or *generalized derivatives* of the function $y = x\sin\frac{1}{x}$ at the point 0; in particular D^+ and D_+ are the upper respectively lower right-hand derivative, D^- and D_- the upper respectively lower left-hand derivative. We can obtain four such numbers at any point of a continuous function, as we will explain later on. To better illustrate this point we will change our previous example a little, so that the function is no longer symmetric at the right and left of the y-axis. We write

$$y = ax\sin\frac{1}{x} \quad \text{at the right}$$

$$y = bx\sin\frac{1}{x} \quad \text{at the left, with } a + b \neq 0.$$

The geometrical representation is then a curve that oscillates on the right between $+ax$ and $-ax$, and between $+bx$ and $-bx$ on the left, and that is continuous in the origin (Fig. 17).

Our four limits of the difference quotient for Δx that becomes indefinitely small are then

$$D^+ = a,\ \ D_+ = -a,\ \ D^- = b,\ \ D_- = -b,$$

[37] that is: we have here an *example in which the four derivates are different.*

We make a further step in generalization.

In fact our example also has the peculiarity that the difference quotient takes on infinitely many times the limit value towards which its upper and lower limits are approaching for Δx converging to zero. But it is easy enough to find examples in which this does not happen.

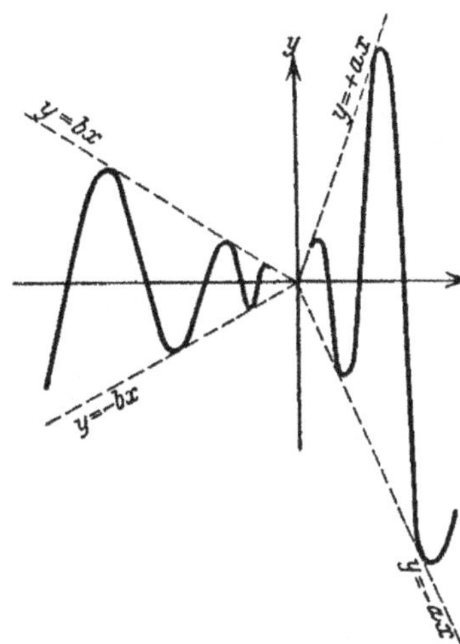

Figure 17

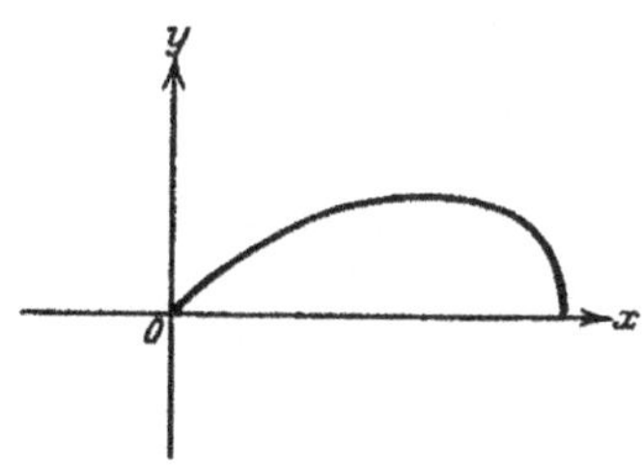

Figure 18

Suppose $y = y_1 \sin \frac{1}{x}$ where y_1 is the ordinate of a curve that, as in Fig. 18, goes through the origin with an inclination of 45°.

In this case, $\frac{\Delta y}{\Delta x}$ never reaches $+1$ or -1, but it comes arbitrarily close to these limit values of its upper und lower limits.

If we choose for y_1 a curve given by a law, for instance a lemniscate (in order to have a closed analytical formula), then y_1 and x are linked by the equation

$$\left(x^2 + y_1^2\right)^2 = x^2 - y_1^2$$

that is, for the part of curve in the first quadrant:

$$y_1 = \left| \sqrt{\frac{\left|\sqrt{8x^2+1}\right| - (2x^2+1)}{2}} \right| .$$

Thus our function is

$$y = \left| \sqrt{\frac{\left|\sqrt{8x^2+1}\right| - (2x^2+1)}{2}} \right| \sin\frac{1}{x}.$$

We obtain for the points $(0,0)$ and (x,y), for any positive x, the following difference quotient

$$\frac{\Delta y}{\Delta x} = \frac{1}{x} \left| \sqrt{\frac{\left|\sqrt{8x^2+1}\right| - (2x^2+1)}{2}} \right| \sin\frac{1}{x}$$

$$= \left| \sqrt{\frac{2\,(1-x^2)}{\left|\sqrt{8x^2+1}\right| + (2x^2+1)}} \right| \sin\frac{1}{x}.$$

For $x \to 0$ the first factor approaches 1, while the second factor oscillates infinitely many times between $+1$ and -1.

Here $\frac{\Delta y}{\Delta x}$ *oscillates between* $D^+ = +1$ *and* $D^- = -1$ *when* Δx *tends to zero, but these values themselves are never taken on by* $\frac{\Delta y}{\Delta x}$.

With these examples the following should be clear: Take any continuous function $y = f(x)$. We mark the point (x, y) and construct on the right Δx and the
[38] corresponding Δy, then there exist two limit values D^+ *and* D_+ to which the upper and the lower limit of $\frac{\Delta y}{\Delta x}$, when Δx becomes arbitrarily small, tend. Analogously, when Δx tends to zero from the left, the limits of $\frac{\Delta y}{\Delta x}$ tend to two determined limit values D^- *and* D_-. The limits can also assume the values $+\infty$ or $-\infty$. Therefore: Whatsoever properties a continuous function $f(x)$ might have, the four limits of the difference quotient $\frac{\Delta y}{\Delta x}$ tend, as Δx tends to zero, always to D^+ *and* D_+ from one side, and to D^- *and* D_- from the other side[49].

If $D^+ = D_+$ we will say that the function has a right derivative; analogously, if $D^- = D_-$, we will say that it has a left derivative.

Only if the four derivates are all equal can we speak of the single specific derivative of a function at a point.

If one clarifies to oneself how many conditions are necessary for a function to have a derivative at each point, one must wonder if functions with such a property actually exist.

This state of amazement corresponds to the other extreme with respect to the thoughtless habitual idea according to which each continuous function should have a derivative[50].

---

[49] [A detailed study of the four derivates can be found in *Constantin Carathéodory*, Vorlesungen über reelle Funktionen, 2nd edition, Leipzig 1927.]

[50] [In the well-known work "Les Atomes" by *Jean Perrin* (translated into German by *Alfred Lottermoser*, 1914) we find an observation that completes perfectly what we have described. Perrin says

Weierstraß's Non-Differentiable Function. Its General Characteristics

After these general remarks we now deal in detail with an *example of a continuous function, which has a derivative in the described sense at none of its points.*

We choose the famous function found by *Weierstaß* around 1861[51], which was **[39]** published only in 1874 in a paper of *Paul du Bois-Reymond* (Journal für reine und angewandte Mathematik, Vol. 79 1875)[52].

The Weierstraß function is defined by an infinite series of the form

$$y = \sum_0^\infty b^\nu \cos(a^\nu \pi x)$$

where $b > 0$*; but, in order for the series to converge, it has to be* $b < 1$*, and* a *and the product* ab *have to satisfy certain conditions, which will be further specified.*

---

on pp. IX and X of his preface: "We completely remain in the experimental reality if we observe with a microscope the Brownian motion of a small particle suspended in a fluid. To put a tangent on its trajectory, we should at least find an approximate limit value for those straight lines that connect the positions of a particle at two close moments. Now, during such investigations, the direction changes incessantly if we decrease more and more the distance between two observations. From these investigations the impartial observer can only imagine a function without derivative, and by no means a curve with a tangent". You can also see the interesting observations of *Emile Borel* in his conference of 1912 "Les théories moléculaires et les mathématiques", printed as Note VII in his "Introduction géométrique à quelques théories physiques" (Paris, 1914).]

[51] Translator's note: strangely, no historical scrutiny has been applied here. As Weierstraß explains in his belated publication in 1895 of his 1872 lecture (Werke, Vol. 2, p. 71), he credits Riemann with having expressed doubts in 1861, or even earlier, that non-differentiability of continuous functions might occur at more than one isolated point. It was Weierstraß himself who publicly announced in 1872 the famous function mentioned here. He allowed du Bois-Reymond to publish it, which occurred in 1875 (not in 1874, as is given erroneously). See Schubring, Gert (2012), Lettres de mathématiciens français à Weierstraß – documents de sa réception en France, *L'aventure de l'analyse, de Fermat à Borel. Mélanges en l'honneur de Christian Gilain*, Éd. Suzanne Féry (Nancy: Presses Universitaires de Nancy), 567–594.

[52] [We can find *Weierstraß's* own description in Vol. 2 of his works, pp. 71–74. We can read the interesting letters of *Weierstraß* to *Paul du Bois-Reymond* and *Leo Koenigsberger*, which are published in vol. 39 (1923), pp. 199–239, of Acta Mathematica. As discovered only a few years ago, 30 years before *Weierstraß, Bolzano* had already constructed an example of a continuous function that is never differentiable. Also see *Gerhard Kowalewski*: Bolzanos Verfahren zur Herstellung einer nirgends differenzierbaren stetigen Funktion. *Leipziger Berichte (mathematisch-physikalische)* Vol. 74 (1922), pp. 91–95; Über Bolzanos nichtdifferenzierbare stetige Funktion. Acta mathematica. Vol. 44 (1923), pp. 315–319. As for the usual references concerning the same argument, we refer to *Konrad Knopp:* Ein einfaches Verfahren zur Bildung stetiger nirgends differenzierbarer Funktionen. *Mathematische Zeitschrift*, Vol. 2 (1918), pp. 1–26; and *Arthur Rosenthal:* Neuere Untersuchungen über Funktionen reeller Veränderlichen. *Enzyklopädie der mathematischen Wissenschaften* Vol. 2, pp. 1091–1096.]

First of all, we want to clarify the structure of the function by means of a *numerical example*. Let $b = \frac{1}{2}, a = 5$; then

$$y = \cos \pi x + \frac{1}{2} \cos 5\pi x + \frac{1}{4} \cos 25\pi x + \frac{1}{8} \cos 125\pi x + \cdots$$

Let us first look at the *partial curves*

$$\begin{aligned} y_0 &= \cos \pi x, \\ y_1 &= \frac{1}{2} \cos 5\pi x, \\ y_2 &= \frac{1}{4} \cos 25\pi x, \\ &\dots\dots\dots\dots\dots\dots \end{aligned}$$

from their superposition we obtain the curve y.

The image of the curve $y_0 = \cos \pi x$ has the usual shape of the cosine with half-wavelength 1, whose first positive zero lies at $x = \frac{1}{2}$.

The function $y_1 = \frac{1}{2} \cos 5\pi x$ yields a wave line whose height oscillates between $+\frac{1}{2}$ and $-\frac{1}{2}$, whose first positive zero lies at $x = \frac{1}{10}$ and whose half-wavelength is $\frac{1}{5}$. The wave line y_1 therefore proceeds *more steeply* than y_0. We can measure the steepness of the partial curves in a suitable way by means of the absolute value of the slope at a zero. For y_0 one obtains as steepness $\left|\frac{dy_0}{dx}\right|_{x=\frac{1}{2}} = \pi$ while for y_1 it has increased already up to $\left|\frac{dy_1}{dx}\right|_{x=\frac{1}{10}} = \frac{5\pi}{2}$.

[40] The function $y_2 = \frac{1}{4} \cos 25\pi x$ describes a sinusoidal line, which proceeds again steeper than the preceding one. Its first positive zero is in $x = \frac{1}{50}$ the half-wavelength is $\frac{1}{25}$ the ordinates oscillate between $+\frac{1}{4}$ and $-\frac{1}{4}$. As for the steepness we obtain $\frac{25\pi}{4}$.

The successive partial curves proceed correspondingly with an always increasing steepness and decreasing height. The amplitudes decrease according to a geometrical series with quotient $\frac{1}{2}$, the wavelengths instead decrease more rapidly, as a geometrical series with quotient $\frac{1}{5}$ while the slopes increase very rapidly according to a series with quotient $\frac{5}{2}$. As a crucial point – and being aware of its essentialness for the eventual success – we note: *The single partial curves, whose superposition composes the final curve, determine waves with decreasing amplitude but with rapidly increasing steepness.*

Now we leave the numerical example and go back to the original function $y = \sum_0^\infty b^\nu \cos a^\nu \pi x$. Here we distinguish between

1. *partial curves* $y_\nu = b^\nu \cos a^\nu \pi x$ and
2. *approximating curves* $Y_m = \sum_0^m b^\nu \cos a^\nu \pi x$.

(Y_m is the sum of $m + 1$ partial curves).

With reference to the partial curves, for $a > 1$, $b < 1$ we have anyhow the following theorem: By increasing ν the amplitude and the wavelength of the partial curves become indefinitely smaller. But do the partial curves also become steeper and steeper, as in our numerical example?

The derivative $\frac{dy_\nu}{dx}$ becomes equal to $-a^\nu b^\nu \pi \sin a^\nu \pi x$ for a zero of y_ν we therefore have

$$\left|\frac{dy_\nu}{dx}\right| = (ab)^\nu \pi.$$

As soon as $ab > 1$ the steepness of the partial curves increases without a bound for increasing ν, although their amplitude decreases indefinitely.

Let us now say something about the *approximating curves.* If we already have understood how such an approximating curve originates from the preceding one, then we will be able to gain somehow a picture of the final curve arising from indefinitely continuing this process.

It is

$$Y_1 = \cos \pi x + b \cos a\pi x,$$

this means, therefore: we must imagine the approximating curve Y_1 as arising from superposing the usual cosine curve with a finer wave-line.

If we construct Y_2, then we obtain **[41]**

$$Y_2 = \cos \pi x + b \cos a\pi x + b^2 \cos a^2\pi x,$$

that is, we add here to the previous curve a still finer one.

Although this way the organic construction of the approximating curves is realised[53], our imaginativeness is rapidly weakening when we want to look at the **[42]** final curve as a whole.

We rather need to be happy with its logical definition, which I formulate as follows:

Each successive approximating curve is obtained from the preceding one by superposing a new finer wave of smaller amplitude but disproportionately smaller wavelength.

What can we say now about the *final curve*?

We consider the m-th approximating curve as constructed and write

$$y = \sum_0^\infty b^\nu \cos a^\nu \pi x = Y_m + \sum_{m+1}^\infty b^\nu \cos a^\nu \pi x.$$

[53] Compare Fig. 19. Here again it is $b = \frac{1}{2}, a = 5$. With this figure and the following observations I agree with the developments made by *Christian Wiener* in Journal für reine und angewandte Mathematik, 90 (1881), pp. 221–252, without accepting his incorrect critique of Weierstraß's assertions. (Wiener's text is very instructive with respect to the unclear ideas that are occasionally connected to the concept of derivative. You can compare this to the clear reply of *Weierstraß* (Works, Vol. 2, pp. 228–230).

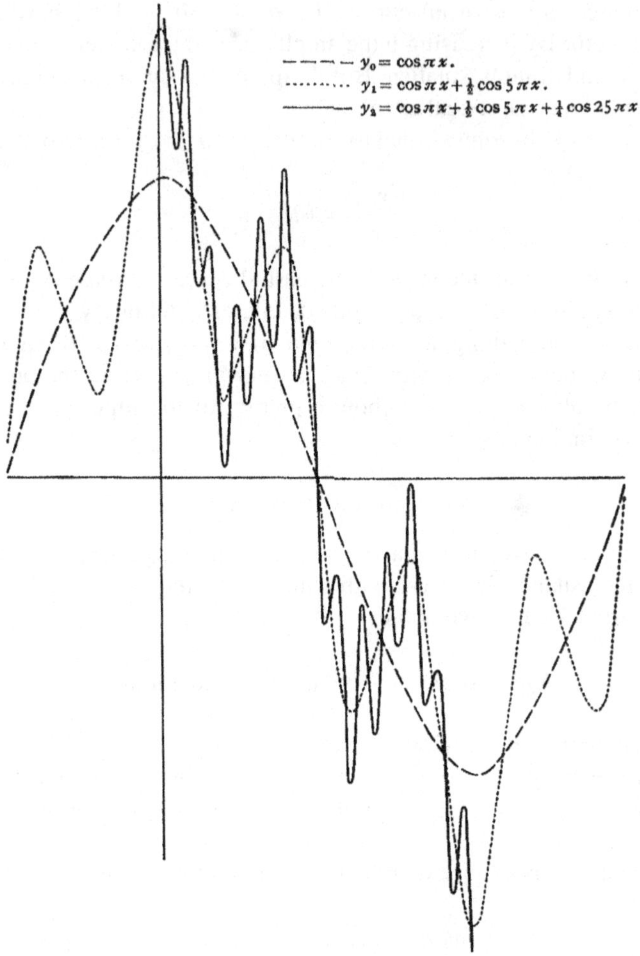

Figure 19 Approximating curves of the Weierstraß function $y = \sum_0^\infty b^\nu \cos a^\nu \pi x$ for $b = \frac{1}{2}, a = 5$

The single addends of the sum $\sum_{m+1}^\infty b^\nu \cos a^\nu \pi x$ all have the same factor b^{m+1}. If we factor it out we have

$$y = Y_m + b^{m+1} \sum_0^\infty b^\nu \cos a^{m+1+\nu} \pi x.$$

Now we are searching for an estimate for the infinite sum and will this way assign an upper and lower bound for y, that is, we determine the stripe in which the Weierstrassian function lies.

We can obtain an upper bound by assigning to the cosines always the value $+1$, and we can obtain a lower bound by assigning them always the value -1. So we obtain

as the upper bound for y: $Y_m + b^{m+1}(1 + b + b^2 + \ldots)$

as the lower bound for y: $Y_m + b^{m+1}(-1 - b - b^2 \ldots)$

In the brackets, we have a geometrical series, which converges because we made the positive b expressly smaller than 1. So the result is $y = Y_m + \varepsilon b^{m+1} \cdot \frac{1}{1-b}$, where ε is an unknown factor between $+1$ and -1, with these limits if need be included. In words we can say:

The ordinate of the final curve is obtained from the ordinate Y_m of the approximating curve of index m by adding a quantity of the form

$$\varepsilon \frac{b^{m+1}}{1-b} \quad \text{with} \quad -1 \leqq \varepsilon \leqq 1.$$

From a more geometrical point of view, we can also say:

Our final curve is contained in a stripe of width $2\frac{b^{m+1}}{1-b}$ whose central line is the approximating curve Y_m.

For instance, if $b = 0{,}1$ and we are measuring in centimetres, for $m = 6$ we obtain the width $2\frac{b^{m+1}}{1-b} = 2\frac{0{,}1^7}{0{,}9} = \frac{2}{9} \cdot 10^{-6}$, that is, a width which is less than what can be perceived by the sharpest microscope. So, if b is fairly small, the width decreases extraordinarily rapidly as m increases. *In any case, the continuity of the function y always follows from this*, as such:

Y_m is continuous being the sum of continuous functions. Whatever is added in order to obtain the final curve is uniformly small for any x when m is sufficiently increased. Considering the two things together, this means nothing else but the continuity of y. We want to briefly give the formulas for this: **[43]**

Assuming

$$y(x) = Y_m + \varepsilon \frac{b^{m+1}}{1-b},$$

$$y\left(x'\right) = Y'_m + \varepsilon' \frac{b^{m+1}}{1-b},$$

we get

$$\left|y(x) - y\left(x'\right)\right| = \left|Y_m - Y'_m + \left(\varepsilon - \varepsilon'\right) \frac{b^{m+1}}{1-b}\right|,$$

$$\left|y(x) - y\left(x'\right)\right| \leqq \underbrace{\left|Y_m - Y'_m\right|}_{\eta_1} + \underbrace{\left|\left(\varepsilon - \varepsilon'\right) \frac{b^{m+1}}{1-b}\right|}_{\eta_2},$$

and therefore

$$\left|y(x) - y\left(x'\right)\right| \leqq \eta_1 + \eta_2 = \eta.$$

Expressed in words this becomes:

We initially take m so great, that we obtain $\left|(\varepsilon-\varepsilon')\cdot\frac{b^{m+1}}{1-b}\right| \leqq 2\frac{b^{m+1}}{1-b} < \frac{\eta}{2}$, then we take x' so near to x, that it results $|Y_m - Y'_m| \leqq \frac{\eta}{2}$; therefore also $|y-y'|$ becomes smaller than any given positive η. We summarize:

Since the width of the stripe can be reduced as much as desired for $b < 1$ by increasing m and since furthermore the approximating curve Y_m is a continuous function, then the final curve is also continuous for $b < 1$. But we can say even more about the final curve. *With a suitable choice of a it is possible to identify, from the first, particular points of the final curve in a closed form, namely its nodes and vertices.*

First the *nodes*:

The cosine of an angle becomes zero if the argument is an odd multiple of $\frac{\pi}{2}$; let us take $x = \frac{2g+1}{2a^m}$ where g is an integer number. Then we have

$$a^m\pi x = (2g+1)\frac{\pi}{2}, \quad \cos a^m\pi x = \cos(2g+1)\frac{\pi}{2} = 0.$$

If we also assume that a is an odd number, then – since the product of two odd numbers is an odd number – also $a^{m+1}\pi x, a^{m+1}\pi x, \ldots$ are of the form $(2g+1)\frac{\pi}{2}$ and we obtain

$$\cos a^{m+1}\pi x = \cos a^{m+2}\pi x = \cdots = 0.$$

Thus, if we assume $x = \frac{2g+1}{2a^m}$ where g is an arbitrary integer number and a an odd number, then the m-th partial curve does not only have a zero in this point, but also each successive partial curve, and the ordinate of the final curve coincides in this point with the $(m-1)$-th approximating curve and all the successive ones.
[44] Therefore, on the final curve there are points that lie simultaneously on the $(m-1)$-th approximating curve and on all the successive ones.

We call these points *nodes*; moreover, they lie on the ordinate lines that correspond to the zeros of the partial curves. If $2g+1$ could be divided by a, then the node in question would already have appeared in a previous approximating curve; but we do not need to consider this case any further.

As for the exponent m and also for g we can take arbitrarily large integer numbers, and having presupposed $a > 1$, it follows – by the way – that the points corresponding to the ordinates of the nodes on the x-axis lie everywhere densely, that is, in every arbitrarily small subinterval there are infinitely many nodes.

The *vertices* are a bit more difficult, but they are of fundamental importance in proving that the Weierstrassian function is nowhere differentiable. We will use this term for certain points of the approximating curves as well as of the final curve. The vertices of Y_m are the points of this curve that we obtain for $x = \frac{g}{a^m}$, g is again an integer number, and the vertices of the final curve are given by the totality of the points corresponding to $x = \frac{g}{a^m}$ $(m = 0, 1, 2, \ldots)$. The m-th partial curve in these points has its maxima and minima, because in these it is:

$$\cos a^m\pi x = \cos\pi g = (-1)^g.$$

Since a is odd, it is also

$$\cos a^{m+1}\pi x = \cos a\pi g = (-1)^g.$$

The same holds for all successive partial curves, so that we obtain the result: The vertices of the approximating curves and consequently of the final curve lie on the ordinate lines that pass through the maxima and minima of the partial curves.

For a vertex in $x = \frac{g}{a^m}$ the ν-th term of the Weierstraß function series has the value $b^\nu(-1)^g$ for $\nu \geqq m$.

As a consequence, we can easily give the sum of the series for a vertex. Let us denote with Y_{m-1} the ordinate that the approximating curve of index $m-1$ has in this point, then we obtain:

$$y = Y_{m-1} + (-1)^g \cdot \frac{b^m}{1-b}.$$

If we project the vertices of the Weierstrassian function on the x-axis, these projections obviously lie everywhere densely as those of the nodes, for the same reason. The result of our reflection is therefore:

When $0 < b < 1$ and a is an odd number greater than 1, the Weierstraß function is everywhere continuous and it determines on the x-axis two everywhere dense point sets known to us. Being a continuous function, it is completely determined by **[45]**
the values assumed at the points of one of these two sets.

Its Non-Differentiability

The question that now follows systematically is:

What happens at every x to the four limits of the difference quotients D^+, D_+, D^-, D_-?

We will find, that the four derivatives do not have at any point a common finite or infinite value, so that our function does not have at any point a finite or infinite derivative. Of course, it suffices to show that two of the four derivatives, for instance D^+ and D_-, have different values.

Weierstraß's proof is not detailed because the points x are no further classified. We will follow his proof in order not to waste too much time with this topic; a more detailed analysis is in any case possible and easy.

Let x_0 be the point at which we want to investigate the existence of the derivative. We look first at the m-th approximating curve. Weierstraß chooses the nearest of the two vertices of this *approximating curve* adjacent to x_0; if x_0 lies exactly in the middle, he chooses the left one (Fig. 20).

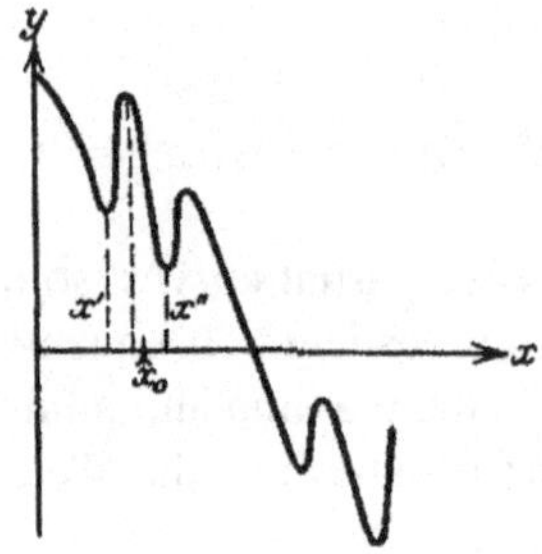

Figure 20

If $\frac{\alpha_m}{a^m}$ is the abscissa of this vertex, where of course α_m is an integer number, we have the inequality

$$-\frac{1}{2} < x_0 a^m - \alpha_m \leqq +\frac{1}{2}.$$

The vertex determined in this way has, in turn, two adjacent vertices on the m-th approximating curve; the one on the left has abscissa x' and the one on the right has abscissa x''. For x' and x'' we determine the ordinates of the *final curve* and we connect with the given point (x_0, y_0) the points (x', y') and (x'', y'') that we have obtained. This way we determine two difference quotients of the final curve and we need to check whether they tend to a common finite or infinite limit value, when the two differences $x' - x_0$ and $x'' - x_0$ both tend to zero. We will see that for sufficiently great values of the product ab

$$\frac{y' - y_0}{x' - x_0} \quad \text{and} \quad \frac{y'' - y_0}{x'' - x_0}$$

tend to unlimitedly great values that either have different signs, or both oscillate between $-\infty$ (as a lower limit) and $+\infty$ (as upper limit), which excludes the existence of a finite or of an infinite derivative.

Let us at first analyse the differences $x' - x_0$ and $x'' - x_0$.

Since

$$x' = \frac{\alpha_m - 1}{a^m} \quad \text{and} \quad x'' = \frac{\alpha_m + 1}{a^m}$$

[46] we obtain

$$x' - x_0 = \frac{\alpha_m - 1 - x_0 a^m}{a^m} = \frac{-1 - (x_0 a^m - \alpha_m)}{a^m},$$

$$x'' - x_0 \qquad = \frac{1 - (x_0 a^m - \alpha_m)}{a^m}$$

or, if we introduce, as Weierstraß did, the notation x_{m+1} for the difference $x_0 a^m - \alpha_m$, we have

$$x' - x_0 = \frac{-1 - x_{m+1}}{a^m}, \quad x'' - x_0 = \frac{1 - x_{m+1}}{a^m}.$$

Since $|x_{m+1}| \leqq \frac{1}{2}$, it follows that for $a > 1$ the two differences tend to zero as m increases; we could have directly derived this from the fact that the point set on the x-axis corresponding to the vertices of the final curve is everywhere dense.

We write the values for y' and y'' as follows

$$y' = \sum_{0}^{m-1}{}_{\nu}\, b^{\nu} \cos a^{\nu}\pi x' - (-1)^{\alpha_m}\, b^m \sum_{0}^{\infty}{}_{\nu}\, b^{\nu},$$

$$y'' = \sum_{0}^{m-1}{}_{\nu}\, b^{\nu} \cos a^{\nu}\pi x'' - (-1)^{\alpha_m}\, b^m \sum_{0}^{\infty}{}_{\nu}\, b^{\nu}$$

If we split up the difference $y' - y_0$ into two parts, we have for the difference quotient $\frac{y'-y_0}{x'-x_0}$

$$\frac{y'-y_0}{x'-x_0} = \sum_{0}^{m-1}{}_{\nu}\, b^{\nu} \frac{\cos a^{\nu}\pi x' - \cos a^{\nu}\pi x_0}{x'-x_0} + \sum_{0}^{\infty}{}_{\nu}\, b^{m+\nu}\, (-1)^{\alpha_{m+1}}\, \frac{1+\cos a^{\nu}\pi x_{m+1}}{x'-x_0}.$$

The idea, which justifies the splitting, is to separate the difference quotients of the $(m-1)$-th approximating curve $\sum_{0}^{m-1} b^{\nu} \cos a^{\nu}\pi x$ and of the "leftover-curve" $\sum_{0}^{\infty} b^{\nu} \cos a^{\nu}\pi x$. Now we pass to an estimation of these two quotients.

For the difference quotient of the $(m-1)$-th approximating curve we obtain, transforming the given difference in the numerator into a product,

$$\sum_{0}^{m-1}{}_{\nu}\, b^{\nu} \frac{\cos a^{\nu}\pi x' - \cos a^{\nu}\pi x_0}{x'-x_0} = \sum_{0}^{m-1}{}_{\nu}\, b^{\nu} \frac{2 \sin a^{\nu}\pi \frac{x_0+x'}{2} \cdot \sin a^{\nu}\pi \frac{x_0-x'}{2}}{x'-x_0}.$$

Multiplying both numerator and denominator by $a^{\nu}\pi$ and writing $\frac{1}{\frac{x_0-x'}{2}}$ instead of $\frac{2}{x_0-x'}$, it follows that

$$\sum_{0}^{m-1}{}_{\nu}\, b^{\nu} \frac{\cos a^{\nu}\pi x' - \cos a^{\nu}\pi x_0}{x'-x_0} = \sum_{0}^{m-1} a^{\nu} b^{\nu} \pi \frac{\sin a^{\nu}\pi \frac{x_0+x'}{2}}{a^{\nu}\pi \frac{x_0-x'}{2}} \cdot \sin a^{\nu}\pi \frac{x_0+x'}{2}.$$

As both $\left|\frac{\sin a^{\nu}\pi \frac{x_0-x'}{2}}{a^{\nu}\pi \frac{x_0-x'}{2}}\right|$ and $\left|\sin a^{\nu}\pi \frac{x_0+x'}{2}\right|$ are ≤ 1, we obtain that the absolute value **[47]** of the difference quotient of the $(m-1)$-th approximating curve is smaller or equal to $\pi \sum_{0}^{m-1}{}_{\nu}\, a^{\nu} b^{\nu} = \pi \frac{a^m b^m - 1}{ab-1}$ and surely it cannot reach the value $\pi \frac{a^m b^m}{ab-1}$ (provided that

ab is greater than 1). We are therefore allowed to assume the difference quotient of the $(m-1)$-th approximating curve equal to

$$\varepsilon\pi\frac{a^m b^m}{ab-1}, \qquad (-1 < \varepsilon < +1).$$

Now we turn to the difference quotient of the leftover-curve and obtain

$$\sum_0^\infty{}_\nu\, b^{m+\nu}(-1)^{\alpha_{m+1}}\frac{1+\cos a^\nu \pi x_{m+1}}{x'-x_0} = (-1)^{\alpha_m} a^m b^m \sum_0^\infty b^\nu \frac{1+\cos a^\nu \pi x_{m+1}}{x_{m+1}+1}$$

The first element of the infinite series occurring here is

$$\frac{1+\cos \pi x_{m+1}}{x_{m+1}+1}.$$

Since

$$-\frac{1}{2} < x_{m+1} \leqq +\frac{1}{2},$$

one has $\cos \pi x_{m+1} \geqq 0$. The denominator $x_{m+1}+1$ oscillates between $\frac{1}{2}$ and $\frac{2}{3}$ Thus we have

$$\frac{1+\cos \pi x_{m+1}}{x_{m+1}+1} \geqq \frac{2}{3}.$$

Likewise, the successive elements of the series are positive or zero, so that the sum of the series is $\geqq \frac{2}{3}$.

Considering a positive number $\eta' \geq 1$, and putting $\frac{(-1)^{\alpha m}}{\eta'} \cdot \varepsilon = \varepsilon'$ (so that ε' lies, as ε, between -1 and $+1$) we can now write

$$\frac{y'-y_0}{x'-x_0} = (-1)^{\alpha_m} a^m b^m \eta' \left(\frac{2}{3} + \varepsilon' \frac{\pi}{ab-1}\right).$$

So we have succeeded in obtaining an estimation of the left-hand difference quotient. We can obtain an analogous formula for the right-hand difference quotient, namely

$$\frac{y''-y_0}{x''-x_0} = (-1)^{\alpha_m+1} a^m b^m \eta'' \left(\frac{2}{3} + \varepsilon'' \frac{\pi}{ab-1}\right),$$

but here we have a new factor (-1) with respect to the preceding formula, because now $x''-x_0$ is positive, while $x'-x_0$ was negative above.

[48] Now we want to obtain that for the difference quotients the contribution given to the difference quotient by the leftover-curve is greater then the contribution given by the $(m-1)$-th approximating curve.

By means of a qualitative consideration we arrive to the conclusion:

We have to take care that the *partial waves* that are superposed to the $(m-1)$-th approximating curve, become *as steep as possible.* But the steepness depends on the product ab; so we need to make ab sufficiently great.

Quantitatively the matter is as follows:

We choose the worst case in which ε' (and also ε'') are equal to -1; in this case it must be $\frac{2}{3} > \frac{\pi}{ab-1}$, that means

$$ab > 1 + \frac{3\pi}{2}.$$

We consider this condition as now satisfied. Then $\eta'\left(\frac{2}{3} + \varepsilon'\frac{\pi}{ab-1}\right)$ is surely a positive number p'_m, that depends on m. The same holds for

$$\eta''\left(\frac{2}{3} + \varepsilon''\frac{\pi}{ab-1}\right) = p''_m$$

We obtain

$$\left.\begin{aligned} p'_m &= \eta' \cdot \frac{2}{3} + \eta' \cdot \varepsilon' \frac{\pi}{ab-1} \\ p''_m &= \eta'' \cdot \frac{2}{3} + \eta'' \cdot \varepsilon'' \frac{\pi}{ab-1} \end{aligned}\right\} \geqq \left(\frac{2}{3} - \frac{\pi}{ab-1}\right).$$

If we shorten this by writing

$$\frac{2}{3} - \frac{\pi}{ab-1} = q,$$

we have

$$\left.\begin{aligned} p'_m \\ p''_m \end{aligned}\right\} \geqq q,$$

where q is a number that does not depend on m.

So the following relations will hold for the difference quotients:

$$1.\ \begin{aligned} \frac{y'-y_0}{x'-x_0} &\geqq (-1)^{\alpha_m} a^m b^m q, \\ \frac{y''-y_0}{x''-x_0} &\leqq (-1)^{\alpha_m+1} a^m b^m q, \end{aligned} \qquad \text{or} \qquad 2.\ \begin{aligned} \frac{y'-y_0}{x'-x_0} &\leqq (-1)^{\alpha_m} a^m b^m q, \\ \frac{y''-y_0}{x''-x_0} &\geqq (-1)^{\alpha_m+1} a^m b^m q, \end{aligned}$$

depending on whether $(-1)^{\alpha_m} = +1$ or $(-1)^{\alpha_m} = -1$.

Thus we obtained the final formulas on which the Weierstrassian proof for the non-existence of a derivative is based. In fact, in our difference quotient we let m increase, so that x' from the left and x'' from the right indefinitely approach x_0. If a finite derivative would exist, then the two difference quotients would unlimitedly approach the same finite value. If a determined *infinite* derivative would exist, both **[49]** the difference quotients would finally increase with the same sign towards infinity, as soon as $|x'-x_0|$ and $|x''-x_0|$ become sufficiently small. But if we let m increase in both of the final formulas for the difference quotients, then we have to distinguish three cases:

1. All α_m (with the exception of a finite number) are even; then

$$\frac{y'-y_0}{x'-x_0} \text{ tends to } +\infty, \qquad \frac{y''-y_0}{x''-x_0} \text{ tends to } -\infty.$$

2. All α_m (with the exception of a finite number) are odd; then

$$\frac{y'-y_0}{x'-x_0} \text{ tends to } -\infty, \qquad \frac{y''-y_0}{x''-x_0} \text{ tends to } +\infty.$$

3. If neither case 1 nor case 2 is given, it does not affect generality to suppose that $(-1)^{\alpha_m}$ has alternate signs as m increases. Then the various $\frac{y'-y_0}{x'-x_0}$ as well as the $\frac{y''-y_0}{x''-x_0}$ form an oscillating series with lower limit $-\infty$ and upper limit $+\infty$. Therefore, the *Weierstrassian* function has in no case a finite or infinite derivative at the point x_0, which was chosen completely arbitrarily.

Figure 21

I want to make the essential point even more understandable with a figure. Let us take (see Fig. 21) x_0 in a lower vertex (α_m odd) of the m-th approximating curve and imagine that it remains a lower vertex for all successive approximating curves. This presupposes that all $\alpha_{m+\nu}(\nu = 0, 1, 2, \ldots)$ are odd. Thus the difference in height of neighbouring vertices and also their horizontal distance from x_0 decreases when m increases, but having chosen such a great product ab, the difference in height decreases disproportionately slower than the horizontal distance. The consequence is, that the two secants corresponding to the difference quotients, become indefinitely steeper as m increases, so that their steepnesses become infinitely great and have opposite signs. Here we assume that a lower vertex of the m-th partial curve still remains a lower vertex for the $(m+1)$-th, $(m+2)$-th, ... approximating curve. If all the numbers $\alpha_{m+\nu}(\nu = 0, 1, 2, \ldots)$ are even, x_0 of course becomes an upper vertex of the respective partial curve and the neighbouring vertices become lower vertices. The figure is therefore inverted, but the contradiction to the assumption of the existence of a determined derivative remains unchanged[54].

[54] [The reader will think that for the non-differentiability of the Weierstrassian function the condition $ab > 1$ is sufficient, instead of the stronger one that we assumed $ab > 1 + \frac{3\pi}{2}$, and that

The "Reasonable" Functions

Our analysis of the Weierstrassian function has shown that the assumption of continuity for a function does not imply the existence of a derivative. If we want a continuous function to have a first, a second, and higher derivatives, then we have to pose this as an explicit condition. **[50]**

If we look back at the developments of this section, we find the following results: we ascribed to the empirical curve certain properties from the start:

1. continuity,
2. existence of a greatest and a smallest value, of a finite number of maxima and minima in a finite interval,
3. existence of a slope and
4. of a curvature etc.

In order to find something analogous in a function of precision mathematics, we had to explicitly presuppose the following properties successively:

1. continuity in a closed interval,
2. finite number of maxima and minima in a closed interval,
3. existence of a derivative and
4. of a second derivative etc.

Thus, from the totality of functions we selected a particular class, which is nevertheless more general than the class of analytic functions, because we did not require the existence of a derivative of any order, not to mention the representability by means of a Taylor series. I call these functions, as does *Carl Gustav Jacobi*, "*reasonable functions*"[55].

Introducing this expression we can say:

From a qualitative point of view (that is according to their nature) we find the properties, which are usually ascribed to an empirical curve, in those functions **[51]** $y = f(x)$ *of precision mathematics that we call reasonable functions.*

At this point we have not yet solved the question of the *quantitative* congruence that we will talk about in the next section.

---

the condition "a is an odd number" is not relevant to the matter. In fact *Godfrey H. Hardy* (*Transactions of the American Mathematical Society* Vol. 17 (1916), pp. 301 to 325) proved that the conditions $0 < b < 1$ and $ab \geq 1$ are sufficient if we intend by non-differentiability that no *finite* derivative exists. But, if the restriction to finite derivatives is dropped, then the condition $ab \geq 1$ is no longer sufficient to ensure the non-differentiability. These facts were already known to *Weierstraß*, as it emerges from his correspondence with *Leo Koenigsberger* (see above p. [39]), but he never published anything about this. The above mentioned conditions, besides $0 < b < 1$, a being an odd number $ab > 1 + \frac{3\pi}{2}$, have the advantage to allow an elementary investigation.]

[55] Later on we will speak in the same sense of "regular" functions or "regular" curves.

III. The Approximated Representation of Functions

Approximation of Empirical Curves by Means of Reasonable Functions

Given an empirical function, we ask ourselves the question: Is it possible to determine a function $y = f(x)$ of precision mathematics, which can approximate – for each x – its ordinate, slope, and curvature with a sufficient accuracy? Since, to the purpose of approximation, we will not use a complicated function but only functions of a simple analytic type, another question immediately follows the preceding one:

How far can we approximate an empirical curve with respect to its entire shape, slope and curvature, by means of simple analytically defined functions?

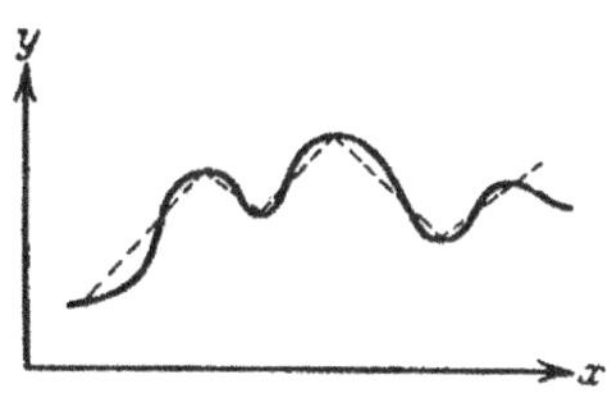

Figure 22

First of all we convince ourselves by an arbitrarily chosen example that each empirical curve can be substituted by a reasonable function with any given degree of approximation. Suppose we need to represent with sufficient accuracy the relation between abscissas and ordinates shown in Fig. 22, by means of a function $y = f(x)$. We inscribe to the curve a rectilinear polygon with such a great number of sides that each side perceptibly coincides with a small part of the curve. We cannot say anything in general on how this can be performed in the particular case: I mean, how long the individual sides must be and where we have to choose the corresponding vertices; in any case we can substitute each empirical curve with a rectilinear polygon with the precision demanded by the circumstances[56]; this

[56] This is a result of practical experience and as such we pose it at the beginning of our mathematical considerations.

F. Klein, *Elementary Mathematics from a Higher Standpoint*,
DOI 10.1007/978-3-662-49439-4_4

polygon acts in place of a law $y = f(x)$ that we can consider as an approximation of the law given by the curve.

However, we have thereby not yet obtained a function that we can differentiate everywhere, let alone obtained a function whose derivative is everywhere coincident with the "slope" of the empirical curve; the correspondence existing within the demanded limits of error refers for now only to the ordinates.

[52] In order to find an analogous result for the slope, the simplest way is to obtain a first "derived curve" from the given empirical curve, whose ordinate is y, by determining for each x of the original curve the difference quotient $\frac{\Delta y}{\Delta x}$ that determines the slope (see p. [24]) and by setting this value as the ordinate y_1 of a new curve $y_1 = f_1(x)$ (Fig. 23).

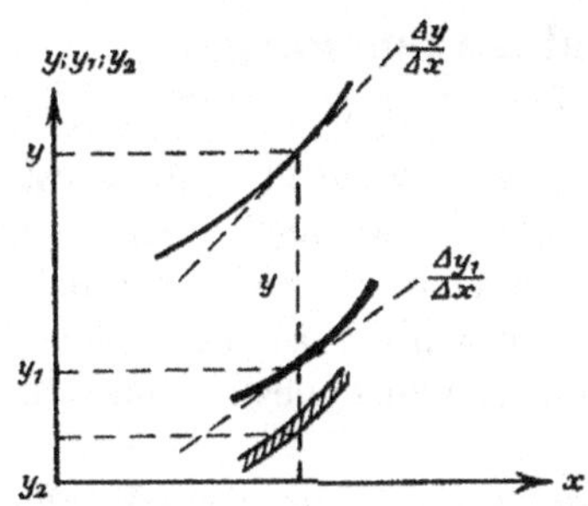

Figure 23

Since the empirical curve is not given in an exact way, but it is represented by a "stripe", the slope $\frac{\Delta y}{\Delta x}$ will oscillate within quite wide precision limits.

The "first derived curve" will thus have a quite significant width; this can be checked by knowing that in any case (within the allowed limits of precision) it must always be $\int f_1(x)dx = f(x)$.

Correspondingly, we can construct a "second derived curve" by setting as ordinate y_2 the slope $\frac{\Delta y_1}{\Delta x}$ of the first derived curve. Of course the latter is then undetermined to a greater extent, etc.

After having carried this out, we substitute the curve y_2 with a polygon and thus obtain a function $f_2(x)$, which, integrated twice, gives a function $f(x)$ that corresponds to the initial curve with respect to ordinates, slopes and curvatures, with the required precision. Briefly in formulas:

$$y_2 = f_2(x), \quad y_1 = \int f_2(x)dx, \quad y = \int y_1 dx = f(x).$$

I summarize again:

In order to determine a twice-differentiable reasonable function in such a manner that it represents not only the ordinates of a given empirical curve but also its slope and curvature within given margins of precision, in correspondence of the given curve we construct the first and second derived curves – as far as this is possible in an empirical domain – substitute the second derived curve for a rectilinear *polygon and this way define a function $f_2(x)$, which – integrated twice – yields the approximating function $f(x)$ we were looking for.*

This is of course only *one possible* method to determine a reasonable function with the desired properties; every mathematician will soon think of different methods.

The details of how to use such a method are subject to practical decisions that vary from case to case. We are not at all solely moving in the logical field of pure mathematics, but rather in a field where – besides the purely logical conclusions of **[53]** mathematics – a certain feeling for what is practical, useful, and accessible is also involved. We do not want to enter particular application fields, but, on the other hand, we do not want to continue to express ourselves in such an undetermined way as we have just done; therefore – in the further question concerning the approximate representation by means of simple analytic formulas – instead of the empirical curve, we want to consider from the start, a "reasonable" function $y = f(x)$, which can, as the case may be, be differentiated once, twice (or even more often), and ask: *How far is it possible to approximate such a reasonable function by simple analytic expressions (polynomials, trigonometric sums, etc.)?*

Therefore we transplant our approximate observation completely into the field of pure mathematics, where all premises can be clearly defined and all statements can be expressed rigorously. But we leave it to the "applied" disciplines to judge what can be practically achieved through our developments in the specific cases.

So we will now treat the

Approximation of a Reasonable Function by Means of Simple Analytic Expressions

The analytic formulas that are most used for approximate representations are

a) finite polynomials

$$y = A + Bx + Cx^2 + \cdots + Kx^n,$$

b) finite trigonometric series (trigonometric polynomials or sums)

$$\begin{aligned} y = a_0 + a_1 \cos x + a_2 \cos 2x + \cdots + a_n \cos nx \\ + b_1 \sin x + b_2 \sin 2x + \cdots + b_n \sin nx. \end{aligned}$$

We have to discuss to what extent reasonable functions can be represented by means of these simple functions and in particular to what extent the approximation also concerns the derivative.

In textbooks, the approximation of functions by finite series is often neglected with respect to the question of the exact representation by infinite series (Taylor's, Fourier's, etc.). This, however, is a completely different question, which is indeed important, but which never comes into play in applications. In fact, according to the nature of applications, the question can only be how far and in what sense can we approximate using *finite* series. The one-sidedness of many textbooks can be only explained by the fact that the authors do not write based on practice but only based on concerns about theoretical considerations.

[54] When approximating we can use one of the following ideas. We can

a) link the finite polynomials and the finite trigonometric series only to individual points of the function or of its derivatives

b) deduce the coefficients from the entire course, for instance reducing to the minimum value the sum of the squares of the errors, according to the method of least squares[57].

I start with polynomials that are linked only to individual points of the function.

Given the function $y = f(x)$, we assign n abscissas $x = \alpha, \beta, \ldots, \nu$ in which the required polynomial has to take on exactly the values of the ordinates of the function.

We obtain the polynomial of lowest degree, which presents only n constants, by means of the well-known

Lagrange's Interpolation Formula

$$Y = f(\alpha)\frac{(x-\beta)(x-\gamma)\cdots(x-\nu)}{(\alpha-\beta)(\alpha-\gamma)\cdots(\alpha-\nu)} + f(\beta)\frac{(x-\alpha)(x-\gamma)\cdots(x-\nu)}{(\beta-\alpha)(\beta-\gamma)\cdots(\beta-\nu)} + \cdots$$
$$+ f(\nu)\frac{(x-\alpha)(x-\beta)\cdots}{(\nu-\alpha)(\nu-\beta)\cdots}.$$

The formula yields a polynomial of degree $(n-1)$, which has in fact in the given abscissas exactly the same ordinates as the function $y = f(x)$. For $x = \beta$ it yields, for instance, $Y = f(\beta)$.

Now we have to investigate to what extent the Lagrange approximation polynomial approximates our function in other abscissas different from the given n points.

To decide this, we call $\Theta(x)$ our polynomial Y and $R(x)$ the remainder $y - Y$. Thus we have

$$y = \Theta(x) + R(x).$$

The expression $R(x)$ vanishes for $x = \alpha, \beta, \ldots$, therefore we can factor out $\varphi(x) = (x-\alpha)(x-\beta)\cdots(x-\nu)$ and write

$$y = \Theta(x) + \varphi(x) \cdot r(x).$$

At the same time we give Lagrange's formula the form

$$\Theta(x) = \frac{f(\alpha)}{\varphi'(\alpha)}\frac{\varphi(x)}{x-\alpha} + \cdots + \frac{f(\nu)}{\varphi'(\nu)}\frac{\varphi(x)}{x-\nu}.$$

---

[57] [These problems have already been treated in volume I (pp. [205]–[215]). Here we omit unnecessary repetitions.]

Should Lagrange's formula be useful in a given interval, then $|r(x)|$ has to remain sufficiently small in the same interval ($\varphi(x)$ is bounded in every closed interval). Therefore we have the following key question: **[55]**

Is it possible to enclose $r(x)$ within certain limits so that $\Theta(x)$ can stand as an approximate expression of $f(x)$?

The field in which we are moving here is usually referred to with the term *interpolation* or interposition. This terminology derives from the fact that the value x, in which we want to approximate the function, is thought to be inside the interval $\alpha, \beta, \ldots, \nu$. But our question also concerns those values x, which are outside the interval (*extrapolation*)[58].

We now pass from this general approach to particular cases in which we suppose that two or more points $\alpha, \beta, \ldots$ are brought to coincide, this means that at particular points we will have the first derivative and also higher ones[59].

If, besides $f(\alpha), f(\beta)\ldots$, we also have $f'(\alpha), f'(\beta)\ldots$ the following question arises:

How can we construct a *polynomial* that in the points $\alpha, \beta, \gamma, \ldots$ has not only the required ordinates, but *also the predetermined derivatives* (osculating interpolation)?

This can be solved either directly or also deduced from Lagrange's formula by a passage to the limit, and the question is, then, to what extent the polynomial thus obtained can be used to approximate the function $f(x)$ and its derivatives.

Taylor's Theorem and Taylor's Series

If, in particular, all points $\alpha, \beta, \ldots, \nu$ are brought to coincide in one point a, we obtain Taylor's formula as follows:

$$f(x) = f(a) + \frac{f'(a)}{1}(x-a) + \frac{f''(a)}{2!}(x-a)^2 + \cdots + \frac{f^{(n-1)}(a)}{(n-1)!}(x-a)^{n-1} + r(x)\cdot(x-a)^n.$$

I do not wish here to enter into the details of how to deduce it from Lagrange's formula, which can be done easily[60].

As to Lagrange's formula and its particular cases, what we are interested in is, as already said, to *estimate the remainder.*

As a basis for this estimate we use the *Theorem of Rolle*, the well-known particular case of the *mean value theorem for differentiation.* The theorem states: *Let $F(z)$ be a continuous function over the closed interval $a \le z \le b$, which has a derivative*

[58] [In volume I, p. [247] it was suggested that it is convenient to use, instead of "interpolation", the term "approximation" which also includes extrapolation. Recently, the term interpolation has tacitly come to include extrapolation].

[59] [More details can be found in Vol. I, pp. [247]–[252].]

[60] [The detailed explanation can be found in Vol. I, p. [247] etc.]

at every point in the interior of the interval. Moreover, let $F(a) = F(b) = 0$. *Then*
[56] *the derivative* $F'(z)$ *has at least one zero in the interior the interval* (Fig. 24).

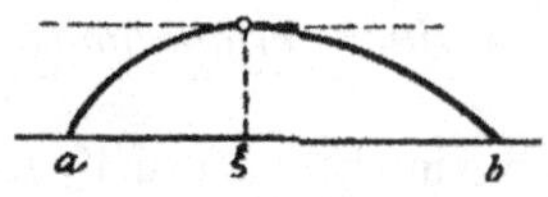

Figure 24

The proof of this theorem is easy. We remember the theorem of *Weierstraß*, which states that every function continuous in a closed interval has either in the interior of the interval or in its endpoints a greatest value. We now exclude the trivial case in which $F(z)$ has a constant value zero and suppose that $F(z)$ is not without positive values inside the interval $a \dots b$. Then $F(z)$ must take on its greatest value in a point ξ in the interior of the interval. From the uniqueness of the derivative it immediately follows that $F'(\xi) = 0$.

Should $F(z)$ take on only negative values inside the interval, then we would look for the greatest value of $-F(z)$ and obtain the same result.

The simple elements of this proof are, as we remarked, the Weierstraß theorem and the assumption of the existence of the first derivative of $F(z)$.

We now extend the property to three points a, b, c, for which $F(z)$ should vanish. Applying our theorem twice we get that $F''(z)$ vanishes at least once in the interval $a \dots c$. So we obtain the following theorem that I express in the form in which we will use it:

If $F(z)$ *is a function that vanishes at k points, and if in the closed interval determined by these zeros* $F(z)$ *is continuous and sufficiently often differentiable, then the* $(k-1)$*-th derivative of* $F(z)$ *has at least one zero in the interior of the interval.*

This theorem leads us to estimate the remainder in Lagrange's interpolation formula as follows.

We consider the function

$$F(z) = f(z) - \Theta(z) - r(x) \cdot \varphi(z),$$

where $\Theta(z)$ (of degree $n-1$) represents *Langrange*'s polynomial without the remainder and $\varphi(z)$ (of degree n) the factor introduced on p. 54; x is an arbitrary, but fixed value of the independent variable z. We now know a series of zeros of the function $F(z)$. First of all, for $\alpha, \beta, \dots$, both $f(z) - \Theta(z)$ and $\varphi(z)$ equal zero, and therefore $F(z) = 0$. Moreover x is a zero, because for $z = x$ we have, according to the definition, $f(x) = \Theta(x) + r(x) \cdot \varphi(x)$; so $F(x) = 0$.

Thus we can apply to $F(z)$ Rolle's theorem in its general form taking $k = n+1$.
[57] It follows, that $F^{(n)}(z)$ has at least one zero ξ inside the interval $\alpha \dots \nu, x$. If we calculate this n-th derivative we obtain

$$F^{(n)}(z) = f^{(n)}(z) - r(x) \cdot n!$$

(being $\Theta(z)$ of degree $(n-1)$, its n-th derivative yields the additive factor 0, being $r(x)$ a multiplicative constant, it gives the value $r(x)$, while the n-th derivative of

$\varphi(z) = z^n + \cdots$ yields the factorial $n!$). So we have for the corresponding ξ:

$$f^{(n)}(\xi) - r(x)n! = 0$$

and therefore

$$r(x) = \frac{f^{(n)}(\xi)}{n!}.$$

Inserting the resulting value of $r(x)$, obtained in a simple but revealing manner, in Lagrange's remainder formula we obtain

$$f(x) = \Theta(x) + \varphi(x)\frac{f^{(n)}(\xi)}{n!}.$$

where ξ, which we do not know otherwise, lies somewhere in the interval $\alpha, \beta, \ldots, \nu, x$.

Thus we have the result:

Lagrange's formula is usable, in other words, $f(x)$ *is approximated by* $\Theta(x)$, *provided that the expression* $\varphi(x) \cdot f^{(n)}(\xi) : n!$ *yields a sufficiently small value for each* ξ *in the interval.* It does not matter whether x lies inside or outside the values $\alpha \ldots \nu$. Our remainder formula serves both for interpolation and extrapolation.

If we now want in particular to write down *Taylor*'s formula, we have:

$$f(x) = f(a) + \frac{f'(a)}{1}(x-a) + \cdots + \frac{f^{(n-1)}(a)}{(n-1)!}(x-a)^{n-1} + \frac{f^{(n)}(\xi)}{n!}(x-a)^n.$$

The form of the remainder here is the same, which one uses in teaching the differential calculus when dealing with Taylor's series with Lagrange's remainder.

By the way, with Taylor's formula we have to do only with extrapolation, as all $\alpha, \beta, \ldots, \nu$ coincide with the same point a.

I now want to clarify with a series of examples the meaning of what we have just exposed.

First of all, I want to speak of the application of Lagrange's formula that all of you use with the table of logarithms. Suppose we find in the table the logarithm of two numbers a and $a+1$. The question is: Can we interpolate linearly, that is, can we substitute the small part of the logarithmic curve between the ordinates a and $a+1$ with the chord? (Fig. 25).

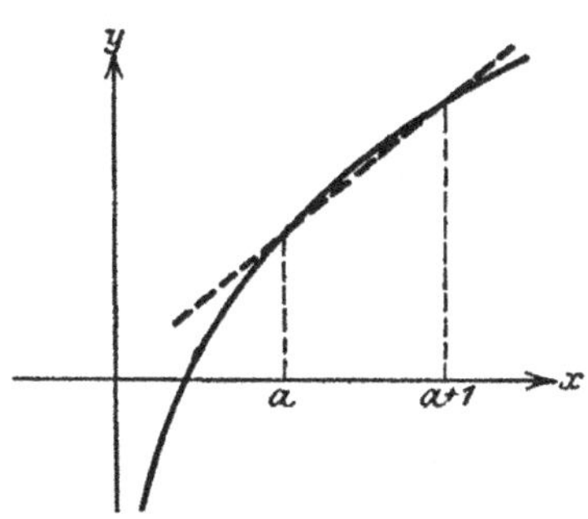

Figure 25

[58] In the case of $n = 2$ and for $f(x) = \log x$ Lagrange's formula yields:

$$\log x = \log a + (x - a)\{\log(a + 1) - \log a\} + R.$$

For the evaluation of the remainder

$$R = \frac{f''(\xi)}{2}(x - a)(x - a - 1),$$

we determine $f'(x) = \frac{d \log x}{dx} = \frac{M}{x}$ and $f''(x) = -\frac{M}{x^2}$ where M is the modulus of the logarithm[61]. The curve representing the product $(x-a)(x-a-1)$ is a parabola that, in the interval $a \leq x \leq a + 1$, takes on its greatest ordinate $-\frac{1}{4}$, in absolute value, for $x = \frac{a+(a+1)}{2} = a + \frac{1}{2}$ (Fig. 26).

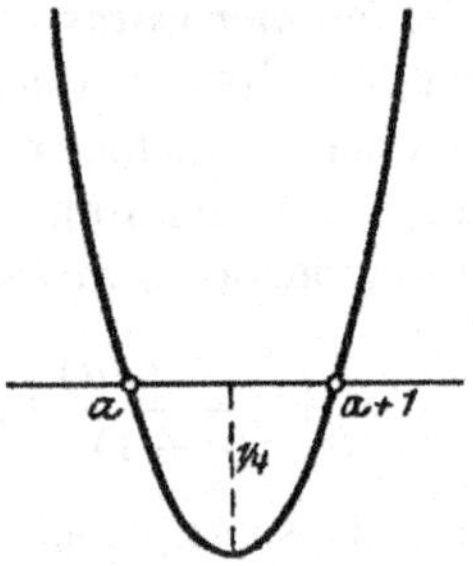

Figure 26

If we enter these values in the formula above, we obtain as an upper bound for the error $|r(x)|$ that we commit by omitting the remainder $\frac{M}{8\xi^2}$. So:

When we calculate intermediate logarithmic values by means of linear proportionality, we obtain a value that is too small, but at most by $\frac{M}{8\xi^2}$. Here ξ is an arbitrary value between a and $(a + 1)$. The absolute value of the error is therefore smaller than $\frac{M}{8a^2}$.

Let us understand the amount of the error numerically. Let us use tables with seven digits, where we can surely find the logarithms of five-digit numbers. So both a and ξ have five digits. We substitute the modulus $M = 0{,}43429\ldots$ of the usual logarithms with the somewhat larger number 0,5 and – when choosing for a the smallest five-digit number (for which the error is the largest) – we obtain as upper bound for the error $\frac{0{,}5}{8 \cdot 10000^2} = \frac{1}{16 \cdot 10^8}$, a value that is sensibly smaller than the error that results from the fact that the logarithms of the five-digit integer numbers are given only up to the seventh digit[62].

---

[61] Translator's note: Usually $M = \log_{10} e = 1/\ln 10 = 0{,}4342\ldots$

[62] [In his last lecture (1911) on differential and integral calculus, Klein presented in detail the calculation of the rounding error, which is due to the fact that in interpolation we do not use the precise values of $\log a$ and $\log(a+1)$ but only approximations. An account of this unfor-

The second example is given to explain *Taylor's theorem*:

$$f(x) = f(a) + \frac{f'(a)}{1!}(x-a) + \cdots + \frac{f^{(n-1)}(a)}{(n-1)!}(x-a)^{n-1} + \frac{f^{(n)}(\xi)}{n!}(x-a)^n .$$

This formula still has its valuable meaning when $f(x)$ is not indefinitely differentiable. However, in every textbook it is usually presupposed that $f(x)$ is indefinitely differentiable. If, by increasing n, the absolute value of the remainder falls below any given value, $f(x)$ can be put equal to the emerging infinite series. This is often the case for sufficiently small $|x-a|$; we then have an "interval of convergence". If the series converges for every finite x, $f(x)$ is called an *entire*[63] function. **[59]**

If there is convergence at all, it is interesting to analyse the behaviour of the approximating curves corresponding to the consecutive partial sums of the Taylor's series. This has been done in detail in the first volume of this work with the help of drawings, so that we do not need to come back to this now[64].

Approximation of the Integral and of the Derivative Using Lagrange's Polynomial

Now I move on to analysing a further question: *How far does Lagrange's formula with remainder enable the approximation of the integral or of the derivative of a function $y = f(x)$?*

We start with the question of the *approximation of the integral.*

The important things can be found in most textbooks, because the numerical evaluation of the area circumscribed by a curve on the basis of Lagrange's formula is used everywhere (the so-called *mechanical quadrature* or *numerical integration*). On the other hand, the evaluation of the error is often omitted or not dealt with in such detail as would be necessary for practical uses of the formula[65]. I give here only a short survey of four important simple cases, without going into the details of the calculation.

tunately not yet published presentation, in which Klein uses in a beautiful way the idea of the function stripe as an illustration, is in *Wilhelm Lorey*, Zur geometrischen Veranschaulichung des Taylorschen Lehrsatzes und ähnlicher Annäherungsformeln, *Zeitschrift für mathematischen und naturwissenschaftlichen Unterricht*, Vol. 43 (1912), pp. 544–556.]

[63] Translator's note: Also called *integral* function.

[64] Vol. I, pp. [241]–[245]

[65] [See, for instance, *Carl Runge* and *Hermann König:* Numerisches Rechnen. Berlin 1924; *Johan F. Steffensen:* Interpolationslaere. Kopenhagen 1925 (English edition: Interpolation. Baltimore 1927; German translation in preparation).] Translator's note: We can not confirm the publication of such German translation.

Suppose we have to calculate $\int_a^b f(x)dx$.

a) The values of the function in points a and b are given. We substitute the part of the curve lying between $f(a)$ and $f(b)$ with the chord (Fig. 27). From Lagrange's formula for $n = 2$, by integration we obtain

$$\int_a^b f(x)dx = \frac{f(a) + f(b)}{2}(b-a) - \frac{f''(\xi)(b-a)^3}{12},$$

that means, $\int_a^b f(x)dx$ equals the area of the trapezium plus a remainder, where ξ is an unknown quantity in the interval $a, \ldots, b$.

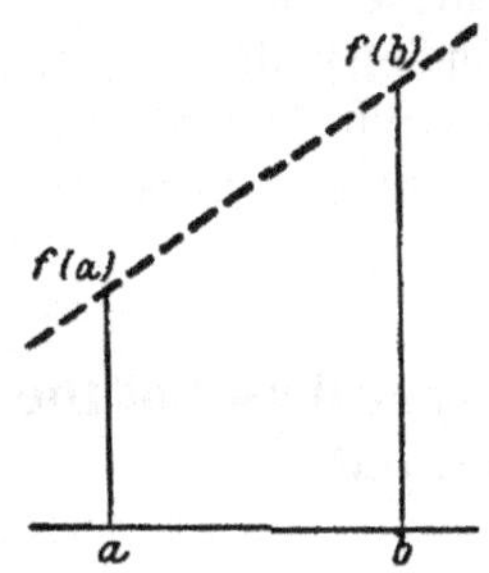

Figure 27

[60] b) another approach to linear interpolation is to substitute the curve with its tangent in the abscissa $\frac{a+b}{2}$. In this case we have

$$\int_a^b f(x)dx = f\left(\frac{a+b}{2}\right)(b-a) + \frac{f''(\xi)(b-a)^3}{24},$$

that means that the error is half the amount of the previous case and it has opposite sign.

What is strange here is that the slope coefficient $f'\left(\frac{a+b}{2}\right)$ does not appear in the final formula. In fact, for the area of the trapezium used to approximate the area of the curve it does not matter in which direction the corresponding side of the trapezium passes through the point $x = \frac{a+b}{2}$, $y = f\left(\frac{a+b}{2}\right)$ (Fig. 28).

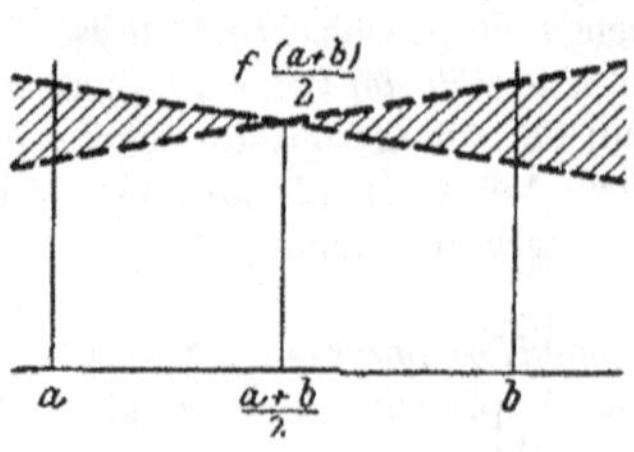

Figure 28

c) If, analogously, we want to determine Lagrange's formula for $n = 4$, that means approximating the curve $y = f(x)$ with a parabola of the third degree, it is convenient to determine the parabola by using for the points a and b both the ordinates and the slopes of the parabola. We obtain

$$\int_a^b f(x)dx = \frac{f(a)+f(b)}{2}(b-a) - \frac{f'(b)-f'(a)}{12}(b-a)^2 + \frac{f^{IV}(\xi)(b-a)^5}{720}.$$

(the simplest case of the so-called *Euler summation formula*).

d) Another way to determine the cubic parabola is to ascribe to the curve the endpoints of the ordinates $f(a)$ and $f(b)$ and the endpoint of the ordinate $f\left(\frac{a+b}{2}\right)$ with given slope $f'\left(\frac{a+b}{2}\right)$ (Fig. 29). Then again $f'\left(\frac{a+b}{2}\right)$ does not appear in the final formula. We obtain

$$\int_a^b f(x)dx = \frac{f(a)+4f\left(\frac{a+b}{2}\right)+f(b)}{6}(b-a) - \frac{f^{IV}(\xi)(b-a)^5}{2880}.$$

The last line shows the determination of the area as it appears in the *so-called Simpson* rule[66].

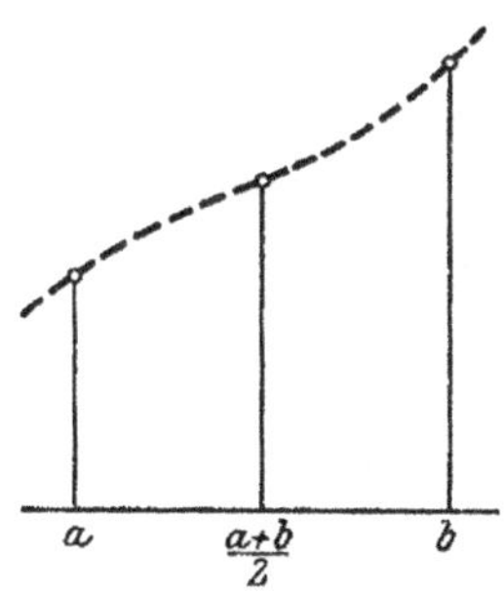

Figure 29

As to the question, to what extent Lagrange's formula with remainder *approximates the derivative of* $f(x)$, we make the following approach:

The function **[61]**

$$f(x) - \Theta(x),$$

has n known zeros $\alpha, \beta, \ldots, \nu$. To compare $f'(x)$ with $\Theta'(x)$ we look for the zeros of the function $f'(x) - \Theta'(x)$.

66 [Among the more extensive works on mechanical quadrature, on the theory of interpolation and numerical calculus we mention, besides *Runge-König* and *Steffensen*, in particular "The calculus of observations" by *Edmund Taylor Whittaker* and *George Robinson* (London 1924).]

From the mean value theorem we deduce that it has at least $(n-1)$ roots $\alpha', \beta', \ldots, \mu'$ between $\alpha, \beta, \ldots, \nu$. I now suggest as approach that – while calling $\psi(x)$ the product $(x-\alpha')\cdots(x-\mu')$ – one sets

$$f'(x) = \Theta'(x) + s(x)\cdot\psi(x)$$

where $s(x)$ can be calculated in an entirely analogous manner as $r(x)$ in the original formula (see pp. [56]–[57]).

It easily follows that

$$s(x) = \frac{f^{(n)}(\xi)}{(n-1)!},$$

so that the result can be expressed in words as follows:

So just like we can put

$$f(x) = \Theta(x) + \frac{f^{(n)}(\xi)}{n!}\cdot\varphi(x),$$

we can as well have an approximation formula for the derivative

$$f'(x) = \Theta'(x) + \frac{f^{(n)}(\xi)}{(n-1)!}\cdot\psi(x),$$

where $\psi(x)$ is the product of the new factors $(x-\alpha'), \ldots, (x-\mu')$, and ξ is an unknown quantity in the interval $\alpha', \ldots \mu', x$.

In particular cases, one clearly has to find out how to deal with the function $\psi(x)$. If the roots of $\psi(x)$ cannot be determined individually, one is nevertheless sometimes able to estimate the value of $\psi(x)$[67].

Analytic Functions and Their Use in the Explanation of Nature

At this point I want to insert some side remarks discussing in general the concept of

analytic function

and clarify to what extent it does or rather does not intervene here.

We start from Taylor's formula

$$f(x) = f(a) + \frac{f'(a)}{1!}(x-a) + \ldots + \frac{f^{(n-1)}(a)}{(n-1)!}(x-a)^{n-1} + \frac{f^{(n)}(\xi)}{n!}(x-a)^n.$$

I explicitly observe once again that Taylor's formula with remainder does not have among its assumptions the unlimited derivability of the function $f(x)$ but only the

[67] [For the interesting mathematical question of numerical differentiation see also the book of Steffensen cited on p. 59.]

n-th derivability. For sake of clarity, I add moreover that, in the applications, the previous formula is, if possible, used in neglecting the remainder in the considered interval. **[62]**

When do we call $f(x)$ an analytic function?

Two conditions are necessary:

First of all, it must be formally possible to continue Taylor's formula till infinity, that means, *the derivatives of arbitrarily high order must exist* and *the remainder must become arbitrarily small if one increases n.*

If these conditions are fulfilled, then, for all points inside a certain interval, we can put

$$f(x) = f(a) + \frac{f'(a)}{1!}(x-a) + \frac{f''(a)}{2!}(x-a)^2 + \cdots \text{ ad inf.}$$

It is convenient to call *Taylor's theorem* the formula with a finite number of addends plus a remainder, and *Taylor's series* the formula with an infinite number of addends, so that we can say:

$f(x)$ is called an analytic function in the neighbourhood of $x = a$, if $f(x)$ can be expanded into a Taylor series inside an interval that includes the point a.

Therewith it is clear that the concept of an analytic function belongs exclusively to precision mathematics. Moreover, we mention the fact that the conditions under which Taylor's series represents a function in an interval have been clarified relatively late.

The older point of view was that to this purpose the only formal convergence of the series was sufficient. Already *Cauchy* remarked that there could be exceptions in single points. But there exist functions which are indefinitely differentiable and that nevertheless cannot be expanded into a Taylor's series in any point. About this issue, *Pringsheim*[68] reached a certain conclusion in 1893, in which he pointed out which conditions had to be added to the indefinite differentiability of the function for the Taylor's series to actually represent the function, that is, so that for unlimitedly growing *n* the remainder falls below any given quantity.

The fact that these different issues are not always distinguished sharply, is surely connected to the point of view we have often mentioned (that can be found in many books), *that in nature it is possible to encounter only analytic functions.*

If we analyse in further depth this point of view, we see that it arises from two sources: **[63]**

1. from the opinion, that in nature we only find functions that are differentiable, and thereby indefinitely differentiable (a conception which seems, in commonplace thinking, been given together with the continuity of the natural phenomena);

[68] Über die nothwendigen und hinreichenden Bedingungen für die Entwickelbarkeit von Functionen einer reellen Variablen nach der Taylor'schen Reihe. In: *Eliakim Hastings Moore* et al. (eds.), Mathematical papers read at the International Mathematical Congress held in Chicago 1893. New York 1896, pp. 288–304, http://link.springer.com/article/10.1007/BF01446971; Über die nothwendigen und hinreichenden Bedingungen des Taylor'schen Lehrsatzes für Functionen einer reellen Variablen, *Mathematische Annalen* Vol. 44 (1894), pp. 57–78 (Original note completed by the translator).

2. from the circumstance that the need for Pringsheim's conditions was not known and nobody cared about such conditions.

To this point of view I obviously oppose a different one: In natural phenomena we only have to do with function stripes and the decision as to whether it is convenient to refer the natural phenomena to particular classes of continuous functions within precision mathematics, goes beyond our perception.

Completely different is the case when mathematicians give preference to analytic functions for mathematical reasons, and I want here to add something about the *mathematical advantages of analytic functions.*

These have their foundation in the fact that an analytic function can also be defined for complex points $x = u + iv$. If we insert the value $x = u + iv$ in the Taylor series we immediately obtain the expansion of the function $f(x)$ for a complex variable. But inversely, the theory of complex functions is limited to the analytic functions that arise in this way. Thus:

Analytic functions of a real variable can be extended to complex variables.

The details of how this is done need to be developed in a lecture dedicated to the theory of functions. Here I will highlight only some salient points.

First of all, let's talk about the *domain of convergence of Taylor's series.*

The theory of functions shows that the domain of convergence of *Taylor*'s series for a point x in the complex plane ($x = u + iv$) is given by the disc having x as its centre and whose border passes through the nearest singular point of the function (Fig. 30). The behaviour of the series *on* the border remains unsettled and has to be decided from case to case.

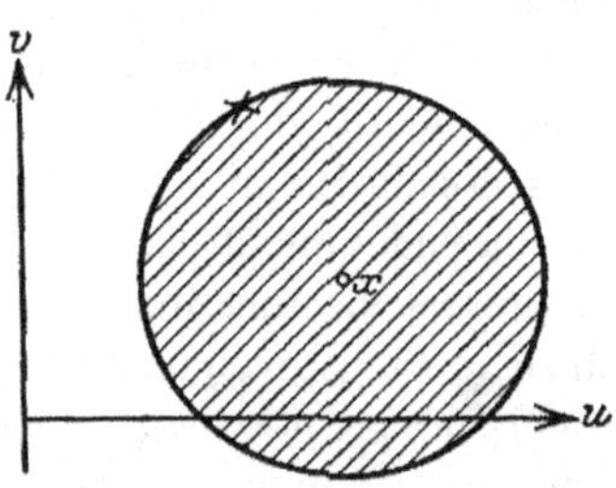

Figure 30

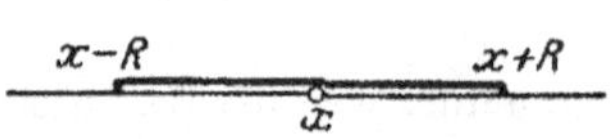

Figure 31

If we limit our observation to real variables, this theorem apparently has a repercussion: The domain of convergence stretches, in the real field, equidistantly to the right and left of the considered point, but how the series behaves in the endpoints remains undecided (Fig. 31).

[64] From the theory of complex functions I further explain the concepts: *function element and analytic continuation.*

Following *Weierstraß*, we call the *Taylor* series of a function $f(x)$ in a point a a *function element of* $f(x)$. It is, thus, defined within the radius of convergence around a.

In many cases we can now go beyond this domain, in that we choose a new value b inside it and transform the original series containing powers of $(x-a)$ into a new series containing powers of $(x-b)$. This new series probably converges in a disc that extends the first one. So one obtains a definition of the function in a new part of the complex plane. This process, which can be continued arbitrarily if desired, is called the *analytic continuation of the function*. Every occurring Taylor's series is called a function element of the function.

To summarize this we can say: The original series directly yields only a first element of the analytic function considered. But in many cases we can define the function in ever increasing domains, extending them by means of the so-called analytic continuation. *The analytic function is in its essence the conception of the first element and of the new function elements that arise from it through analytic continuation.*

A more detailed presentation would be a topic of a lecture about analytic functions. We are interested in the philosophical side of the issue. Our first question is: What do we have to know about an analytic function in order to find a function element of it (from which we can then further deduce the general course of the function)?

Obviously it is necessary, but also sufficient, that we know the coefficients $f(a), f'(a), \ldots$ of Taylor's expansion in the point a. These are defined as limit values of difference quotients, so that they are determined when we know the part of the curve belonging to an arbitrarily small neighbourhood of $x = a$. Herewith we have the result: *In an analytic function, the function element is, in turn, completely determined by an arbitrarily small part of the curve* $y = f(x)$[69].

Now it becomes clear what constitutes the particularity of analytic functions. They are already given by an arbitrary small part of their entire course, but obvi- **[65]**
ously only in the sense of precision mathematics, because only in this field it is possible to define the coefficients $f(a), f'(a), \ldots$ as limits of difference quotients.

Let us now look again at the *mathematical representation of natural phenomena*; we generally assume that the coordinates *x, y, z* can be written as analytic functions of time:

$$x = \varphi(t), \quad y = \psi(t), \quad z = \chi(t).$$

This leads us, according to our previous observations, to a *complete determinism*; the trajectory of the particle in an arbitrarily small time-interval steadfastly determines the trajectory in all the following time-intervals.

[69] [For a long time it was believed that only analytic functions have the property that their entire course is determined by the initial value and the corresponding derivatives. However *Emile Borel* has proved that also non-analytic functions with the same property exist. Also see *Torston Carleman, Les Fonctions Quasi Analytiques*. Paris 1926.]

Therefore we must say:

If it is true that in nature there exist only analytic functions, and that in particular the motion of each single particle is ruled by analytic functions $x = \varphi(t), y = \psi(t), z = \chi(t)$, than it is necessarily true that the motion of the World is predetermined by its behaviour in an arbitrarily small time-interval.

These are conclusions, which are unpleasant for some people. Therefore there is a different point of view opposed to this one, according to which the trajectories are not determined by analytic functions, but by *differential equations with analytic coefficients*. Thus, for instance, one obtains

$$\frac{d^2x}{dt^2} = \text{analytic function of } x, y, z, t, \frac{dx}{dt}, \frac{dy}{dt}, \frac{dz}{dt}.$$

This conception arose from the study of theoretical mechanics. It does not lead to the same rigid conclusions as above. In fact, there are solutions of such differential equations with analytic coefficients that present, so to say, "ramification points" where it is not clear whether the point follows one trajectory or another. We can think of the relation between the general integral curves of a differential equation of the first degree and the singular solutions that arise from the envelopes of the first one (Fig. 32).

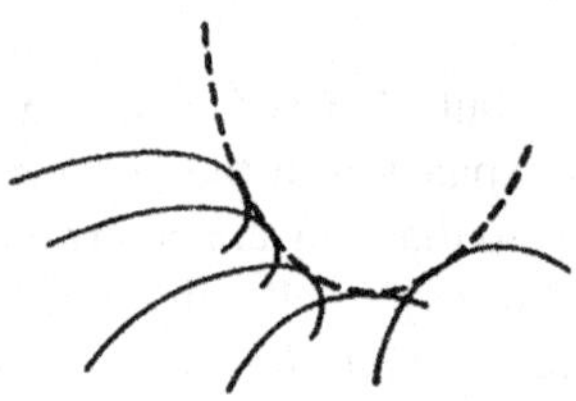

Figure 32

According to the latter point of view, the motion of a point in such a ramification point would not be determined only by the general mechanic laws of nature, but there exists the possibility that the point be determined also by an extra-mechanic cause. In particular, *Joseph Boussinesq* in Paris has passionately developed this idea and he even believes to have found – precisely in the fact that the integration of
[66] differential equations with analytic coefficients can lead to such ramification points – the place in which another power (of biological or ethic nature) can interfere with the general mechanical order of the world.

In his opinion, the occurrence of such places can explain both the existence of free will and the difference between living and dead matter. So, according to him, the principle of the living organism decides, in such a ramification point, the trajectory of the corresponding mass point. On this you can also refer to the publication:

Joseph Boussinesq: Conciliation du véritable déterminisme mécanique avec l'existence de la vie et la liberté morale. Paris: Gauthier-Villars 1879.

You can understand what kind of criticism of this conception I will exert. Boussinesq in all his work treats ideas of precision mathematics (determination of an analytic function through a small part of the function itself etc.). But when we remember that all observations of processes in nature are possible only within limited precision, we see that there is nothing that can justify the assumption that nature itself obeys the considered relations of precision mathematics. If ever precision mathematics should be pertinent, then we remember that this also includes other ideas, for instance that of discontinuous functions of discontinuous variables, which could also be apt to the mathematical representation of nature[70]. But all this discussion, in which observation does not lead us to any decision, occurs from the beginning within the metaphysical field, beyond experience. Thereafter, Boussinesq's theory is not mathematically unsound because he deduces wrong or unsound mathematical consequences from the assumption that in nature we have to do with analytic differential equations, but because his theory assumes as basis the above mentioned unproved statements. We can summarize by saying:

The weak point in all such reflections is that they depend on an arbitrary preference of certain ideas and concepts of precision mathematics, while observations in nature always have only limited precision and can be related in very different manners to topics of precision mathematics[71] . *It is more generally questionable whether we should be looking for the essence of a correct explanation of nature on the basis of precision mathematics, and whether we could ever go beyond a skilful use of approximation mathematics.*

All these discussions had been tied in with Lagrange's interpolation formula, **[67]** which led us to Taylor's theorem and then to Taylor's series and thus to the concept of analytic function. Now we pass over to the

Interpolation by Means of Trigonometric Series

and consider at first finite and then infinite series of this kind.

Let us suppose $f(x)$ to be periodic with period 2π, so that it is $f(x + 2\pi) = f(x)$.

For an approximate representation of this kind of function one prefers as approach the following trigonometric series:

$$\begin{aligned} f(x) &= \frac{a_0}{2} + a_1 \cos x + \cdots + a_n \cos nx + b_1 \sin x + \cdots + b_n \sin nx + R \\ &= \frac{a_0}{2} + \sum_{\nu=1}^{n} (a_\nu \cos \nu x + b_\nu \sin \nu x) + R = \Theta(x) + R, \end{aligned}$$

where the remainder R is omitted if its absolute value is insignificant. The series $\Theta(x)$ contains an odd number $(2n + 1)$ of constants.

---

[70] [See the remarks of *Emile Borel* in his lecture about molecular theories and mathematics, cited on p. [38]]

[71] Compare this to the idea of a completely discontinuous world, in the manner of the cinematograph.

To determine the series $\Theta(x)$ we proceed analogously to the parabolic interpolation and consider at first the $2n + 1$ constants as fixed by $2n + 1$ values $f(x_0), f(x_1), \dots$ of the function $f(x)$. The task is then to establish an expression $\Theta(x)$ that yields for $x = x_0, x_1, \dots$ the function values $f(x_0), f(x_1), \dots$

We can write the required $\Theta(x)$ directly in a form that is analogous to Lagrange's formula, namely in the following way

$$\begin{aligned}\Theta(x) = f(x_0)\,&\frac{\sin\frac{x-x_1}{2}\cdot\sin\frac{x-x_2}{2}\cdots\sin\frac{x-x_{2n}}{2}}{\sin\frac{x_0-x_1}{2}\cdot\sin\frac{x_0-x_2}{2}\cdots\sin\frac{x_0-x_{2n}}{2}}\\ +f(x_1)\,&\frac{\sin\frac{x-x_0}{2}\cdot\sin\frac{x-x_2}{2}\cdots\sin\frac{x-x_{2n}}{2}}{\sin\frac{x_1-x_0}{2}\cdot\sin\frac{x_1-x_2}{2}\cdots\sin\frac{x_1-x_{2n}}{2}}\\ +&\cdots\cdots\cdots\cdots\cdots\cdots\\ +f(x_{2n})\,&\frac{\sin\frac{x-x_0}{2}\cdot\sin\frac{x-x_1}{2}\cdots\sin\frac{x-x_{2n-1}}{2}}{\sin\frac{x_{2n}-x_0}{2}\cdot\sin\frac{x_{2n}-x_1}{2}\cdots\sin\frac{x_{2n}-x_{2n-1}}{2}}.\end{aligned}$$

By repeated application of the known formula for $\cos\alpha \pm \cos\beta$ it is possible
[68] to form from each numerator a sum of cosines and sines of multiples of the angle, which was in fact our original aim. We can retain therefore:

The expression $\Theta(x)$ can be established in complete analogy with Lagrange's polynomial. Thus, only a formal transformation is necessary in order to obtain the form

$$\Theta(x) = \frac{a_0}{2} + \sum_{\nu=1}^{n} (a_\nu \cos \nu x + b_\nu \sin \nu x)$$

This kind of interpolation is often used in practice; but mostly under the particular assumption that all the points $x_0, x_1, \dots x_{2n}$ are *equidistant*, and precisely so that the interval $x_0, x_0 + 2\pi$ is divided into $(2n + 1)$ equal parts. In this case the series $\Theta(x)$ can be calculated with a *direct method*, in an easier way than the one we have just seen. The equidistant points are:

$$x_0,\ x_1 = x_0 + \frac{2\pi}{2n+1},\quad x_2 = x_0 + 2\frac{2\pi}{2n+1}, \dots,\quad x_{2n} = x_0 + 2n\frac{2\pi}{2n+1},$$

let the corresponding function values be denoted by $y_0, y_1, \dots, y_{2n}$. Then, to calculate the coefficients a_μ, b_μ, we have the following $(2n + 1)$ linear equations:

$$y_0 = \frac{a_0}{2} + a_1 \cos x_0 + \cdots + a_n \cos n x_0 + b_1 \sin x_0 + \cdots + b_n \sin n x_0,$$

$$\cdots\cdots\cdots\cdots\cdots\cdots\cdots\cdots$$

$$y_{2n} = \frac{a_0}{2} + a_1 \cos x_{2n} + \cdots + a_n \cos n x_{2n} + b_1 \sin x_{2n} + \cdots + b_n \sin n x_{2n}.$$

The coefficients a_μ, b_μ, can be simply calculated multiplying once each ν-th line of the system by $\cos \mu x_\nu$ and then by $\sin \mu x_\nu (\nu = 0, 1, 2, \ldots, 2n)$, and summing them then up. If one remembers that x_ν and $x_{\nu+1}$ differ by $\frac{2\pi}{2n+1}$ we obtain as a final result – omitting the details:

$$\sum_0^{2n}{}_\nu \, y_\nu \cos \mu x_\nu = a_\mu \cdot \frac{2n+1}{2},$$

$$\sum_0^{2n}{}_\nu \, y_\nu \sin \mu x_\nu = b_\mu \cdot \frac{2n+1}{2}.$$

The interpolation formula thus obtained is used very often in practice, just like the other one, which requires an even number of equidistant observations[72]. If, for instance, in the theory of Earth magnetism we have the problem of determining the **[69]** dependence of the magnetic declination on a latitude circle from the geographical longitude, then one approximately observes the declination[73], for instance at $2n + 1$ equidistant points, calculates the coefficients a_μ, b_μ with the previous formula and obtains the required function in the form of a finite trigonometric series. The remainder is neglected there as being insignificant; whether this is possible to justify has to be judged by the practitioner.

It is now particularly interesting to choose – in the preceding formula -the points, for which the ordinates are given, to lie infinitely dense. Then, from a purely formal point of view, an infinite trigonometric series results for $\Theta(x)$

$$\Theta(x) = \frac{a_0}{2} + \sum_{\mu=1}^{\infty} \left(a_\mu \cos \mu x + b_\mu \cos \mu x\right).$$

We want to see what happens with our formulas for a_μ, b_μ. We denote with Δx_ν the distance $\frac{2\pi}{2n+1}$ between two consecutive ordinates $f(x_\nu)$ and $f(x_{\nu+1})$. From the previous formulas we thus obtain:

$$a_\mu = \frac{1}{\pi} \sum_{\nu=0}^{2n} y_\nu \cos \mu x \cdot \Delta x_\nu,$$

$$b_\mu = \frac{1}{\pi} \sum_{\nu=0}^{2n} y_\nu \sin \mu x \cdot \Delta x_\nu,$$

[72] In meteorology this formula was named after *Friedrich Wilhelm Bessel*. See his paper in *Astronomische Nachrichten* of 1828, pp. 333–348: *Über die Bestimmung des Gesetzes einer periodischen Erscheinung.*

[73] This declination is otherwise provided by the previous interpolation from observations in neighbouring places.

and when we let Δx_ν tend to zero, that is we let n increase without bound, as a limit we obtain the two integrals

$$a_\mu = \frac{1}{\pi}\int_0^{2\pi} f(x)\cos\mu x dx,$$

$$b_\mu = \frac{1}{\pi}\int_0^{2\pi} f(x)\sin\mu x dx.$$

The infinite series $\Theta(x)$ formed with these coefficients is the well-known *Fourier's series*. We postpone the analysis of the remainder $R(x)$ and sum up what we have seen till now in the following way:

The approximation of $f(x)$ by means of an infinite Fourier series is a particular case of approximation by means of a finite trigonometric series over equidistant ordinates.[74]

[74] [See, on this topic, a paper published in *Abhandlungen aus dem Mathematischen Seminar der Hamburger Universität* by *Alwin Walther*: Fastperiodische Folgen und Potenzreihen mit fastperiodischen Koeffizienten (vol. 6: 1, 217–234), where the purely periodic case is treated in reference to *semi-periodic functions*] (Note of 1928 completed by the translator).

IV. Further Discussion About the Trigonometric Representation of Functions [70]

Evaluation of the Error in the Representation of Empirical Functions

With the preceding explanations we have not yet said anything about the remainder, and therefore the convergence, of the Fourier series, whereupon I am going to speak now.

I observe at first that, as far as I know, we cannot find, in the literature, an easy formula for the *remainder of the finite trigonometric series that makes use of finite equidistant ordinates*[75]. Yet, it is this series[76] that one uses the most in applications when one is dealing with periodic phenomena that are to be represented as functions of time. I mention here some disciplines, which make use of these means: Meteorology (representation of the change of temperature and of other meteorological phenomena as functions of time), sound analysis, theory of geomagnetism (dependence of the magnetic elements on the circles of latitude from the geographical longitude), and electrical engineering (change in the intensity of alternating currents over time). It is in fact a result coming from meteorological practice that suggests paying particular attention to estimating the remainder; I want to describe it here in some detail, because it fits the basic ideas of these lectures.

We think of the 24 hours of the day as abscissas on an axis, and, suppose that for four of them, for instance at 12 p.m., 6 a.m., 12 a.m., 6 p.m. we consider the ordinates representing the observed temperature, (Fig. 33).

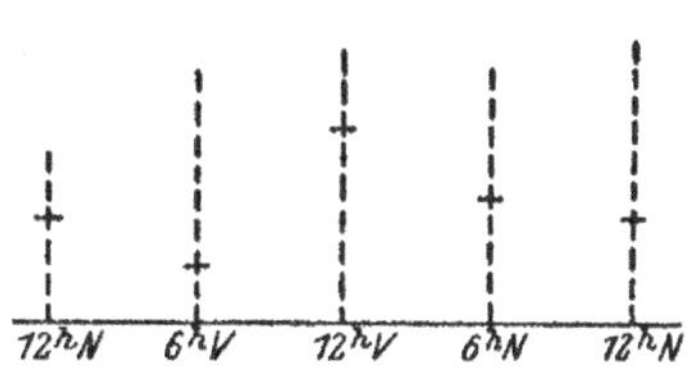

Figure 33

[75] Also see the suggestion in note 79 on p. 72.

[76] Resp. the other one, which we obtain introducing an even number of equidistant ordinates.

F. Klein, *Elementary Mathematics from a Higher Standpoint*,
DOI 10.1007/978-3-662-49439-4_5

Attempts have been made to determine the course of the curve of the temperature through these four values by means of a finite trigonometric series. The curious result (while relying on the conviction that the remainder could be neglected as insignificant) was that the temperature's minimum turns out to be various hours before sunrise. This result, which was presented in all textbooks and that, by the way, gave rise to the strangest explanations, proved to be wrong when the automatic recorders were introduced. These show that the real minimum is immediately before sunrise (for instance at 6 a.m.), in that the curve decreases until the y-value corresponding to sunrise and then increases suddenly again, soon after that (Fig. 34, below). But
[71] this downward motion and the rapid rise of the curve do not appear in the interpolation of the available data, so that the interpolating curve evidently does not give any information about them (Fig. 34, above).

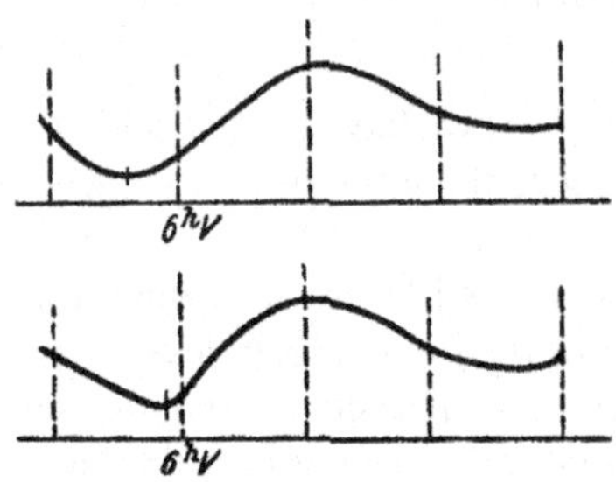

Figure 34

As already mentioned, this considerable mistake is due to the fact that the remainder[77] is being disregarded. So we reach the following result:

In practice, it is *necessary* to control in every single case how many steps are needed of such interpolating formulas in order to obtain sufficient conformity with the observations. An example of the fact that a purely formal interpolation can lead, without such a control, to completely wrong statements is the afore mentioned case of meteorology, which by the way was corrected by the meteorologists themselves.

As to what concerns in particular the "*Fourier*" *series*, the study of its convergence and of its power of representation has become a common property of all text-books and lectures since the classical work of *Lejeune Dirichlet* in volume 4 of Crelle's Journal (1829).[78] The study is mostly carried forth by adding together the first $(2n + 1)$ terms of the series and forming the sum $S_n(x)$, which can be written,

[77] As for literature and details see the report of *Adolph Schmidt*: Über die Verwendung trigonometrischer Reihen in der Meteorologie. Program of the Gymnasium Ernestinum in Gotha of the year 1894.

[78] Translator's note: *Johann Peter Gustav Lejeune Dirichlet*, (1829): Sur la convergence des séries trigonométriques qui servent à représenter une fonction arbitraire entre des limites données. In: *Journal für die reine und angewandte Mathematik (Crelle's Journal)* 4, S. 157–169.

as we derived in Vol. I on p. [211], in the following form:

$$S_n(x) = \frac{1}{2\pi} \int_{-\pi}^{+\pi} f(\xi) \frac{\sin \frac{2n+1}{2} (\xi - x)}{\sin \frac{1}{2} (\xi - x)} d\xi.$$

The integration variable is here denoted with ξ to distinguish it from the value x that has to be considered as fixed. Now one can study the limit to which the integral tends if n increases indefinitely. Indeed, we can see that in many cases the limit is the function $f(x)$ that has to be represented. I leave out the details of the calculation and recall the following:

The sum of the first $(2n + 1)$ terms of the series can be represented as an integral in closed form, and we can convince ourselves that for $n \to \infty$ this integral tends to our 2π periodic function $f(x)$, if, in the interval $0 \leqq x \leqq 2\pi$, the latter satisfies the following so-called *Dirichlet conditions*:

1. $f(x)$ is univocal and bounded.
2. $f(x)$ is piecewise continuous, that is it presents only a finite number of discontinuities. **[72]**
3. $f(x)$ is piecewise monotonic, that is, it has only a finite number of maxima and minima.

These sufficient, but surely not necessary conditions are always fulfilled by a function that is reasonable according to our terminology, in particular when $f(x)$ is "piecewise smooth" for $0 \leqq x \leqq 2\pi$. This is how we call $f(x)$ if the interval $0 \leqq x \leqq 2\pi$ can be divided into a finite number of subintervals so that $f(x)$ itself and its derivative $f'(x)$ are continuous inside each subinterval and approach finite boundary values when tending towards the endpoints from the interior of the subinterval. If the curve $y = f(x)$ is called piecewise smooth, it is in general continuous and has a continuous tangent, but it can present a finite number of jumps and corners (but no cusps with vertical tangent).

By the way, Dirichlet's integral for $S_n(x)$ can also be used to see to what extent a given function $f(x)$ is approximated by a finite number of terms of its Fourier series. You only have to properly estimate this integral[79] (see Vol. I, p. [211] and on).

I also recall here the following properties of the Fourier series:

It has the power to also represent *simple discontinuous functions*, that is, functions with simple jumps (which, like the maxima and minima, are finite in number within $0 \leqq x \leqq 2\pi$). Wherever it jumps, the value given by the series equals the arithmetic mean of those values, which are obtained when approaching the point of discontinuity from the left and from the right (Fig. 35), or, in Dirichlet's notation:

$$\lim_{n\to\infty} S_n(x) = \frac{f(x+0) + f(x-0)}{2}.$$

[79] I suggest the same method for the finite series that makes use of equidistant ordinates.

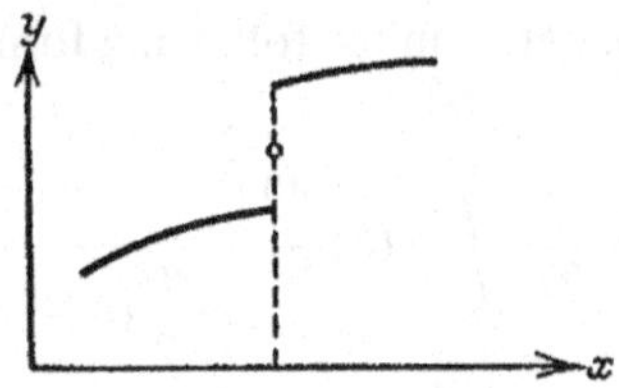

Figure 35

Trigonometric Interpolation According to the Least Squares Method

All these things have already been treated in Volume I of this work. But for the formulas a_μ, b_μ of the coefficients of the Fourier series a different derivation was given, which was first introduced by *Bessel* in his paper mentioned on p. 68. Since we will use later, on a different occasion, the basic ideas of Bessel's derivation, we deem it useful to recall its main features.

When a function $f(x)$ has to be represented by a finite trigonometric series

$$\frac{a_0}{2} + a_1 \cos x + \cdots + a_n \cos nx \\ + b_1 \sin x + \cdots + b_n \sin nx$$

[73] then I can ask myself, roughly speaking, how to determine the coefficients a_μ and b_μ so that the approximation is as "good as possible". It is natural to approach the problem with a method that is similar to that of the *least squares*.

We form the difference

$$\left.\begin{array}{l} f(x) - \dfrac{a_0}{2} - a_1 \cos x - \cdots - a_n \cos nx \\ -b_1 \sin x - \cdots - b_n \sin nx \end{array}\right\} = f(x) - S_n(x),$$

and, considering it as the *error of the representation*, we ask for the minimum of the sum of the squares of the errors over the interval 2π, that is the minimum of the integral

$$\int_0^{2\pi} [f(x) - S_n(x)]^2 \, dx,$$

With this plausible approach we find what we will call the "best" approximation of the function $f(x)$ by a finite trigonometric series. *If we solve this minimum problem we find, independently of how big we choose the value of n, exactly Fourier's values of the coefficients* a_μ and b_μ. It is not even necessary to write down all terms from index 1 until $n (n \geqq \mu)$; on the contrary, we can omit arbitrarily many of them for this approach; the result is always the same for the terms taken into consideration. I conclude with the following theorem:

Any finite collection of terms of the finite Fourier series yields an approximation of $f(x)$, that, from the point of view of the least squares method, is the best of all approximations, that can be achieved with trigonometric terms chosen in the manner described.

The Harmonic Analyser

There exist numerous devices, which allow the mechanical calculation of a certain number of coefficients of the Fourier series, when $y = f(x)$ is given as a drawn curve. These devices are called "*harmonic analysers*" in accordance with the English terminology. In fact, in English the decomposition of a function into the sum of basic periodic terms is called "harmonic analysis", where the word harmonic may remind us that it has been in acoustics that the decomposition of oscillations into individual oscillations of the most simple form has imposed itself for the first time. The analyser that I will present here, in the construction of *Gottlieb Coradi* (Zurich), based on ideas of *Olaus Henrici* (London), yields for each $\mu = 1, 2, \ldots, 6$ an a_μ and b_μ; as for the constant term a_0, it is equal to $\frac{1}{\pi}\int_0^{2\pi} f(x)dx$, and therefore to an area that appears in the initial figure and can be immediately calculated with a planimeter.

Before describing the device in detail, I start with following observation: **[74]**

To calculate the coefficients of the Fourier series, we need to evaluate the integrals

$$a_\mu = \int_0^{2\pi} \frac{f(x)\cos\mu x}{\pi} dx \quad \text{and} \quad b_\mu = \int_0^{2\pi} \frac{f(x)\sin\mu x}{\pi} dx$$

(where the image of $f(x)$ extends somehow from $x = 0$ to $x = 2\pi$). Obviously, we will be able to calculate these integrals in general with the above-mentioned methods of mechanical quadrature and we will be able to do this even when we only have a discontinuous number of observations available[80].

In the textbook

Ernst Gustav Kirsch: Bewegung der Wärme in den Zylinderwandungen der Dampfmaschine. Leipzig 1886,

graphical methods are developed for the approximate calculation of the coefficients[81]. If, on the contrary, the curve is plotted, then we can use a continuous device.

---

[80] The Fourier series then possibly reverts to the above-discussed trigonometric series, with its coefficients defined by sums.

[81] [Also see *Horst von Sanden:* Praktische Analysis 2nd ed. (1924), pp. 122 to 135, where, in addition, the numerical methods are developed. On the execution of the calculation see the already mentioned (p. 59) text of *Carl Runge* and *Hermann König* and the tables of *Leo Wenzel Pollack:* Rechentafeln zur harmonischen Analyse. Leipzig 1926. Moreover, we refer to the papers of *Hans Friesecke, Jan Groeneveld, Walther Lohmann, Richard von Mises, Hilda Pollaczek-Geiringer* and

To understand how our device works it is necessary to transform our integral by means of partial integration. We have

$$a_\mu = \frac{1}{\pi}\int_0^{2\pi} y\cos\mu x dx = \frac{1}{\pi}\left[\frac{y\sin\mu x}{\mu}\right]_0^{2\pi} - \frac{1}{\pi}\int_0^{2\pi}\frac{\sin\mu x}{\mu}dy,$$

$$b_\mu = \frac{1}{\pi}\int_0^{2\pi} y\sin\mu x dx = -\frac{1}{\pi}\left[\frac{y\cos\mu x}{\mu}\right]_0^{2\pi} + \frac{1}{\pi}\int_0^{2\pi}\frac{\cos\mu x}{\mu}dy.$$

Substituting the limits, if $f(x)$ is – not only piecewise – continuous, the first terms on the right hand side fall away, so that the essential integrals, which will be used by the device, are the following:

$$\int_0^{2\pi}\sin\mu x dy \quad \text{and} \quad \int_0^{2\pi}\cos\mu x dy$$

[75] This is how the device manages the evaluation of these integrals:

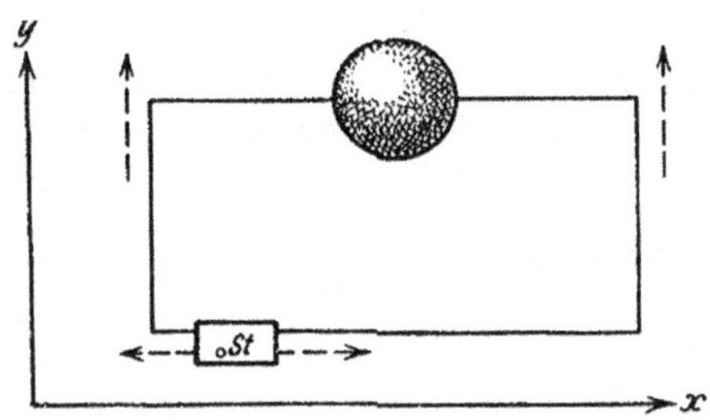

Figure 36

1. First of all, it mainly consists of a right-angled frame, which can be moved only parallel to the y-axis thanks to suitable pulleys, (Fig. 36).

2. Along the frame a slider, moves parallel to the x-axis with a stylus St, so that

3. through the combined motion of the frame and of the slider with respect to the frame, it is possible to move along any given curve with the stylus.

4. The most important component of the device is a satin glass ball on the frame side that is opposite to the slider, which is positioned so that – rolling on the plane when the frame is moved – it undergoes a rotation about a diameter parallel to the x-axis, proportional to dy (Fig. 37).

Ludwig Zipperer in Vol. 2 (1922), Vol. 3 (1923) and Vol. 6 (1926) of the *Zeitschrift für angewandte Mathematik und Mechanik*.]

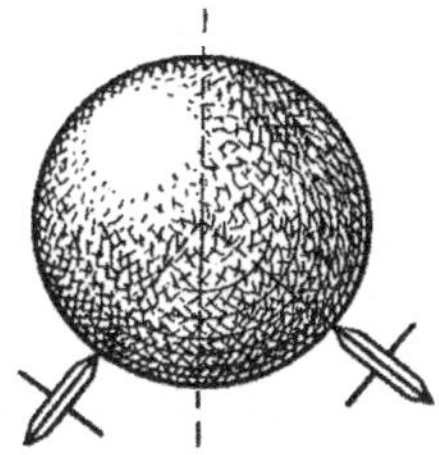

Figure 37

5. If we press a recording wheel on the glass sphere, this will record – from the latitude line against which it presses – a rotation of the ball which is proportional to $dy \cos\varphi$, if φ is the degree of latitude (the proportionality factor depends on the dimensions of the device and is denoted by c). Another equal wheel, inclined 90 degrees with respect to the first one, records a rotation $cdy \sin\varphi$. Both wheels are provided with counters, so that we can immediately read

$$\int c \cos\varphi dy \quad \text{and} \quad \int c \sin\varphi dy$$

6. A further important component of the device ensures that the degrees of latitude of one or the other recording wheels become equal to $\mu\, x$ ($\mu = 1, 2, \ldots, 6$ in our device), as it is necessary to obtain the integrals

$$\int \cos\mu x dy \quad \text{and} \quad \int \sin\mu x dy$$

7. To this purpose, brass disks, with a suitable diameter, are installed on the device above the sphere. These disks are fixed so that their rotation entrains the turning of the two recording wheels. This rotation becomes proportional to the x of the slider because a thin silver wire is connected to it and girdled round the brass **[76]** disk. See Fig. 38.

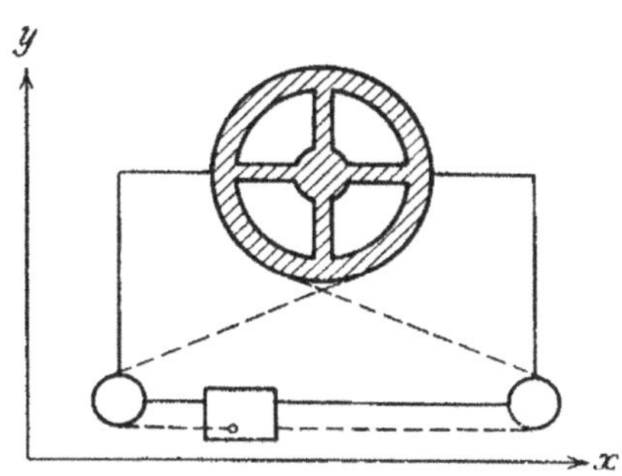

Figure 38

8. If I now choose the radii of the brass disks so that they behave as $1 : \frac{1}{2} : \cdots : \frac{1}{6}$, then the rotation of the disk, and thus the degree of latitude φ of my recording wheels, will be sequentially proportional to $1 \cdot x, 2 \cdot x, \ldots, 6 \cdot x$, and I will be

able to read respectively on both wheels the two integrals, on which it depends for $\mu = 1, 2 \ldots, 6$, multiplied by certain device-constants. The theory is therefore very simple. The actual mathematical invention lies in the transformation of the integral by partial integration. These integrals, thus transformed, become then constructed systematically by means of the device, first by producing dy, then the components $\cos \mu x dy, \sin \mu x dy$, where the main task consists in equalling the degrees of latitude of the recording wheel to the respective μx.

Now, we want to face the question as to how a function is approximated by partial sums of its Fourier series. In that, we mainly rely on drawings that we sketch in connection with two examples.

I am more willingly to confine myself to such an inductive approach for a first introduction, because in this way many details, which are often difficult to understand in the abstract theory, can be immediately understood.

Examples of Trigonometric Series

My first example is a function $f(x)$, whose image is a polygonal path that passes through the points $\pm k\pi (k = 0, 1, \ldots)$ of the x-axis which alternately increase and decrease, each under an angle of 45°. So, the function $f(x)$ is periodical, with period 2π and it takes the following course in the interval $0 \leqq x \leqq 2\pi$:

$$f(x) = x, \qquad \text{if} \quad 0 \leqq x \leqq \frac{\pi}{2},$$

$$f(x) = \pi - x, \qquad \text{if} \quad \frac{\pi}{2} \leqq x \leqq \frac{3\pi}{2},$$

$$f(x) = -(2\pi - x), \qquad \text{if} \quad \frac{3\pi}{2} \leqq x \leqq 2\pi.$$

The function is always continuous, but it is only piecewise analytic, because its first derivative has jump discontinuities for $(2k-1)\frac{\pi}{2} (k = 0, \pm 1, \pm 2 \ldots)$. The Fourier coefficients a_μ of $f(x)$ must all equal zero, because – as we can see immediately –
[77] $f(x)$ is an odd function. So the cosine terms of the series fall away. The coefficients b_μ of the cosine terms can be calculated with the formula given before (compare p. [69]) and we obtain

$$f(x) = \frac{4}{\pi}\left(\frac{\sin x}{1^2} - \frac{\sin 3x}{3^2} + \frac{\sin 5x}{5^2} - \cdots\right),$$

where the equal sign is justified by the fulfilment of Dirichlet's conditions.

In Fig. 39 the curves of the first two approximating functions are represented

$$s_1(x) = \frac{4}{\pi}\frac{\sin x}{1^2},$$

$$s_2(x) = \frac{4}{\pi}\left(\frac{\sin x}{1^2} - \frac{\sin 3x}{3^2}\right)$$

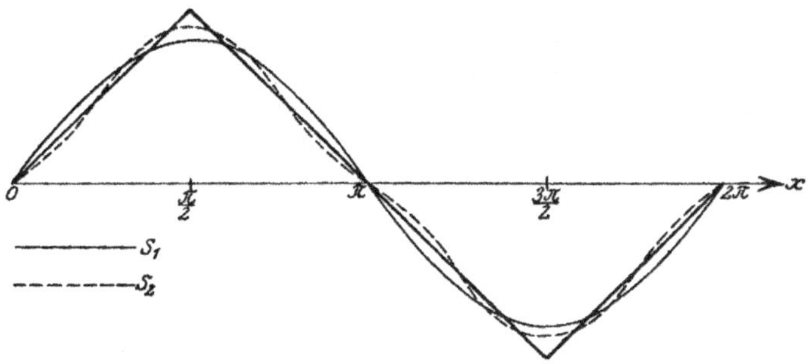

Figure 39

The first approximating curve is like a sine curve, which cuts once the given curve in the interval $\left(0, \frac{\pi}{2}\right)$ and snuggles as near as possible to the triangle situated above $(0, \pi)$; the second approximating curve presents two intersections with the original curve in the interval $\left(0, \frac{\pi}{2}\right)$ and winds along with it oscillating up and down. You might ask *whether by increasing n the number of intersections with the original curve of the approximating curves $s_n(x)$ will increase indefinitely*[82]. In fact this is the case for our example $f(x)$, as we mention incidentally.

As for the approximation of the ordinates of $f(x)$, it improves indefinitely as n increases, and this holds for every x. On the contrary, we cannot expect that for every point also the slope of the original curve will be approximated, in particular this is surely not the case for the points $x = (2k-1)\frac{\pi}{2} (k = 0, \pm 1, \ldots)$. In these points all the approximating functions $s_n(x)$ have derivative equal to zero, that is the curves have a tangent parallel to the x-axis, while the original curve here has a corner.

Before addressing the question of the slope for the other x values, let us consider **[78]**
as a second example a function $g(x)$ with period 2π, which is defined in the interval $0 \leqq x \leqq 2\pi$ as follows:

$$g(x) = \frac{\pi}{4} \qquad \text{if} \qquad 0 < x < \pi,$$

$$g(x) = 0, \qquad \text{if} \qquad x = 0 \text{ or } = \pi \text{ or } = 2\pi,$$

$$g(x) = -\frac{\pi}{4}, \qquad \text{if} \qquad \pi < x < 2\pi.$$

In Fig. 40 the image of $g(x)$ is drawn. Like $f(x)$ in the previous example, also $g(x)$ is piecewise analytic, but, differently from $f(x)$, it is also only piecewise continuous. For Dirichlet's theorem, $g(x)$ can be represented by its Fourier series. One obtains

$$g(x) = \frac{\sin x}{1} + \frac{\sin 3x}{3} + \frac{\sin 5x}{5} + \cdots.$$

[82] [This question is analysed in the paper of *Leopold (Lipot) Fejér*: Über die Fouriersche Reihe, *Mathematische Annalen* Vol. 64 (1907), 273–288.]

In Fig. 41 the curves of the approximating functions $s_1(x)$, $s_6(x)$, $s_{11}(x)$, $s_{16}(x)$ are drawn in a different scale with respect to Fig. 40. In accordance with the previous example, we first observe that here too the number of intersections of the original and the approximating curve increases the more terms of the series are considered[83].

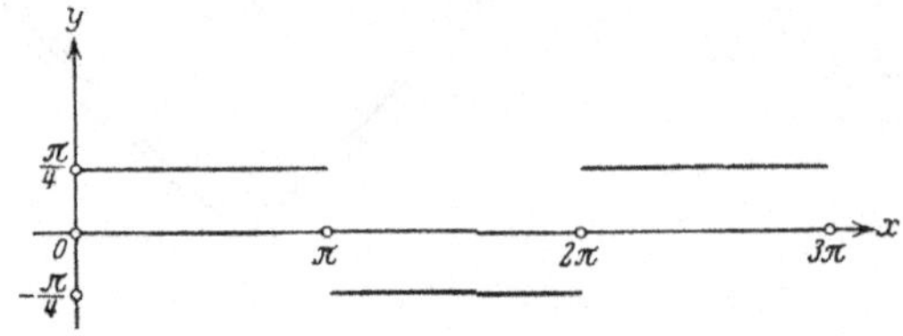

Figure 40

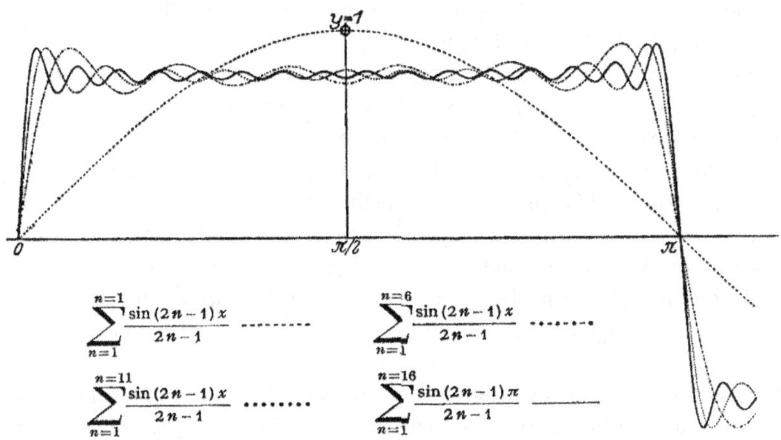

Figure 41

How is it, however, with the approximation of the ordinate itself?

[79] We now arrive to a subtle point, which leads to the so-called "*non-uniform convergence*" of the Fourier series in the neighbourhood of the jump discontinuities of the function that we want to represent. The question is the following:

Let us observe, for instance, the approximating curve $s_6\ (x)$; we will see that it approximates quite well the ordinates of $g\ (x)$ for all x that are not too close to a jump. But if we get sufficiently close to such a point, the approximation gets worse and worse. For instance, if we approach the point 0 from the right – but excluding this point – then the ordinate of the original curve always has the value $\frac{\pi}{4}$, while the ordinate of the approximating curve shoots up from a certain point above the value $\frac{\pi}{4}$, while eventually sinking rapidly down to 0. These facts do not change if we pass to the approximating curves $s_{11}\ (x)$ and $s_{16}\ (x)$. If, keeping n fixed, we approach the

[83] [To produce Fig. 41 we used a slide of the Mathematical Institute of the University Göttingen.]

point of discontinuity getting just close enough, we do not approximate $f(x)$. To summarize this in the form of a general sentence, which per se does not have the value of a proof – but renders the question clearer – I could say:

For any particular value $x = x_0$ we can go so far in the number of terms of the series that the ordinate of the approximating curve matches in a "satisfactory" way the ordinate of the original curve. However, the closer x_0 gets to the point of discontinuity, the greater we have to choose n in order to obtain the desired match. Moreover – considering the n-th approximating curve – x_0 can come so close to the discontinuity point that with this given n no satisfactory match can be achieved. What is inductively explained by the example can be theoretically expressed like this:

The infinite Fourier series of a function $f(x)$ that satisfies Dirichlet's conditions indeed converges to $f(x)$ for all points of continuity x, but when approaching a discontinuity point of the function the convergence appears unlimitedly slowed down[84].

One may think that the series does not converge at all at the points of discontinuity. *But it is exactly the contrary*. This is because the mean value $\frac{g(0+0)+g(0-0)}{2}$ of the limits from the right and from the left, that we are considering, is equal to zero and this value is already given by the first approximating curve. The misconception mentioned is due to the fact that we involuntarily think of the degree of convergence – because of an elementary habit – as a continuous function of x, which is not in fact the case. The phenomenon of the unlimitedly slowed down convergence when approaching a particular point is called "*non uniform convergence*"; the main aim **[80]**
of the given explanation was to clarify this concept.

We already pointed out that the approximating curve $s_6(x)$ and also the successive ones extend quite beyond the value of the original curve when close to a point of discontinuity, before sinking or rising steeply to zero. Following this behaviour of the approximating curves for ever more increasing n, we observe the so-called *Gibbs Phenomenon* already treated in Vol. 1, p. [214] to [215]. It consists in the following: Let AB be the ordinate segment bridging the jump discontinuity. Then, for increasing n, the curves $s_n(x)$ do not approximate AB, but an ordinate CD that extends equally beyond the points A and B. There is

$$AC = BD = \frac{g(x+0) - g(x-0)}{\pi} \cdot \int_{\pi}^{\infty} \frac{\sin x}{x} dx.$$

The value of the integral $\int_{\pi}^{\infty} \frac{\sin x}{x} dx$ is about 0,09 π, so that AC and BD amount to about 9 % of the jump AB. This result holds for all functions $g(x)$ that are represented by their Fourier series and present discontinuities.

If we once again look at Fig. 41, now observing the difference of phase between the maxima and the minima of the different approximating curves, we reach the idea

[84] [See more on this topic in Vol 1., pp. [211]–[213]]. Transl. note: The notion of "indefinitely slow convergence" had been introduced in 1847 by George Stokes.

of using for the approximation, instead of the partial sums $s_n(x)$, their arithmetic mean, and therefore

$$S_0(x) = s_0(x); \quad S_1(x) = \frac{s_0(x) + s_1(x)}{2}; \quad S_2(x) = \frac{s_0(x) + s_1(x) + s_2(x)}{3}; \dots$$

These arithmetic means have been explored by *Feiér* in various well-known papers in the *Mathematische Annalen*[85]. Fejér shows *that with the arithmetic mean $S_n(x)$, the approximation is possible for a larger set of functions then with the partial sums $s_n(x)$ of the corresponding Fourier series.* In many cases, when the Fourier series itself diverges, the arithmetic means of the partial sums converge to the value of the function that has to be represented. *In particular,* the *Gibbs Phenomenon does not appear when approximating by $S_n(x)$* functions of the type presented in Fig. 41.

Now I turn again to the *approximation of the slope* using the approximating functions. In relation to our first curve I want to observe that – with the exception of the vertices – the slope is nevertheless approximated by the successive approximating curves. However, in the second example, the course of the ordinates is indeed
[81] approximated, but not that of the slope. The relative oscillations between approximating curves and the original curve become, on the contrary, steeper and steeper as m increases. Such a behaviour of the approximating curves always occurs when the original curve presents points of discontinuity.

Our example makes us easily understand the reason for such a behaviour.

We can easily draw the derivative of our first curve; it is made up of segments whose ordinates are $+1$ and -1, which alternate in the intervals $\left(0, \frac{\pi}{2}\right)$; $\left(\frac{\pi}{2}, \frac{3\pi}{2}\right) \dots$

In our second example the question is more difficult. In fact, here in general we obtain $g(x) = 0$, and only at the points of jump discontinuities the derivative tends to infinity.

In the first case, the derivative itself can be again represented by a Fourier series

$$f'(x) = \frac{4}{\pi}\left(\cos x - \frac{\cos 3x}{3} + \frac{\cos 5x}{5} - \cdots\right)$$

where at a point of jump discontinuity the mean value of the two values of the jump will be represented. And, as we can notice from the comparison with the Fourier series of the original curve

$$f(x) = \frac{4}{\pi}\left(\frac{\sin x}{1^2} - \frac{\sin 3x}{3^2} + \frac{\sin 5x}{5^2} - \cdots\right)$$

it can almost be obtained differentiating term by term the latter series. So it is clear that our successive approximating curves approximate not only the ordinates, but also – with the exception of the vertices – the slope of the original curve.

---

85 [*Mathematische Annalen*, Vol. 58 (1904), pp. 51–69 and the paper already mentioned on p. [77].]

With our second example the question is completely different. If we want to differentiate the original series term by term

$$g(x) = \frac{\sin x}{1} + \frac{\sin 3x}{3} + \cdots$$

in order to obtain the series

$$\cos x + \cos 3x + \cos 5x + \cdots$$

we see that the latter does not converge, so that we cannot be surprised that the approximating curves do not approximate the derivative. And what we described in our example always happens if we represent functions with jump discontinuities by means of a Fourier series, and consider the derivatives of these functions[86].

Completely different is the behaviour of Fejér's mean of the divergent series: **[82]**

$$\cos x + \cos 3x + \cdots$$

These have – with the exception of the points of discontinuity – for all x the derivative of $g(x)$ as their limit function. The arithmetic means of the original series thus approximate in each interval that does not contain a discontinuity both the ordinate and the direction of the original function $g(x)$. This result is valid for every $g(x)$ that is continuous in the interval $0 \leqq x \leqq 2\pi$ and has a continuous derivative $g'(x)$ (which is piecewise smooth), with the exception of a finite number of places where they have a simple discontinuity. The proof may be read in the aforementioned work of *Fejér* in the Annalen (Vol. 58, 1904).

Chebyshev's[87] Works on Interpolation

These are the few remarks I wanted to make about the Fourier series in this context[88]. At the end of this chapter about interpolation and approximation of functions of a real variable I particularly want to point out the work of *Pafnuty Lvovich Chebyshev*, the great Russian mathematician.

All his life, *Chebyshev* has again and again treated the question of the approximated representation of functions by simple analytic expressions; he has been an *approximation mathematician* par excellence.

---

[86] [On the approximation of the slope through approximating curves of the Fourier series you can find more accurate discussions in Leopold Kronecker, *Vorlesungen über Mathematik*, Vol. 1 (1894), pp. 98–99 and above all in Ernest W. Hobson: The theory of functions of a real variable. 2nd Ed. Cambridge 1926, Vol. 2, pp. 639–643.]

[87] In 19-early 20 century the dominating way of transcription was *Tschebyscheff*, which is the way used by Klein.

[88] [As for the literature concerning Fourier series, I have still to mention the reports for the Enzyklopädie: by *Heinrich Burkhardt* (II A 12) and by *Emil Hilb* and *Marcel Riesz* (II C 10); moreover we cite the presentations in *Courant-Hilbert*: Mathematische Physik I, Berlin 1924, and in *Konrad Knopp*: Unendliche Reihen (already cited on p. 4).]

His treatises have been published, in a French translation, by *Andreevich Markoff* and *Nikolai Sonin* in two volumes. The first volume contains the most varied approaches to the theory of interpolation, and I emphasize three treatises in particular:

1. Sur les fractions continues. *Journal de mathématiques pures et appliquées* (2nd series) Vol. 3 (1858) = Œuvres 1, pp. 203–230;

2. Sur les fonctions qui diffèrent le moins possible de zéro. *Journal de mathématiques pures et appliquées* (2nd series) Vol.19 (1874) = Œuvres 2, pp.189–215;

3. Sur le développement des fonctions a une seule variable. Petersburg, *Bulletin de l'Académie des Sciences*. Vol. 1 (1859) = Œuvres 1, pp. 501–508.

Moreover, I want to mention the following, in order to characterize Chebyshev's research: in our discussion on approximation two questions intersected each other.

Once, we gave the number of the ordinates for a curve and asked for a simple expression that has the same ordinates at the given points. Later, we referred to the ideas of the method of least squares.

[83] These ideas can be combined in the following way: We want to represent a curve through a polynomial of degree n:

$$a + bx + cx^2 + \cdots + kx^n,$$

that has $(n + 1)$ constants. If we only have $(n + 1)$ observations (ordinates) we necessarily come back to Lagrange's interpolation formula. But if we have more than $(n + 1)$ observations, it is possible to use the ideas of least squares, namely by requiring that the sum of the squares of the errors in applying the parabola of n-th degree takes on its minimum. This task can further be generalized, when the different observations have different weight, so that we can try to minimize the sum of $(error)^2 \cdot (weight)$. To this purpose *Chebyshev* gave conclusive formulas thanks to an expansion into continued fractions, so that I can perhaps characterize his achievements briefly as following:

A first series of his works concerns the case in which there are more observations than unknown coefficients to be used in the interpolation formula, and where these coefficients are determined by the requirement that the sum of the squares of the errors, multiplied with appropriate coefficients, has to be minimized.

But *Chebyshev* was not caught by an exclusive faith in the method of least squares, and in further works he obtains that – not the sum of the squares of the errors – but *the absolute value of the greatest errors becomes a minimum.* This is shown in the wonderful paper entitled: *Sur les fonctions qui diffèrent le moins possible de zéro.* You will soon see what is meant. Suppose given a polynomial of degree n whose highest term is x^n; the task is to determine the coefficients of the other terms in such a way that for x lying between $+1$ and -1 the occurring deviations from zero are as small as possible. Here too, *Chebyshev* obtains a very simple final formula.

And now, a third kind of studies.

They refer to the case in which all the ordinates of a curve are considered as given, but with a weight that depends on the abscissa x. He looks for series expansions that well approximate such a given curve according to the principles of the least squares method, wherein he finds all the series used in practice, depending on the distribution of the weight of the different observations. A practical example of *Chebyshev* refers to the straight-line motion obtained via a linkage (the law of the linkage is the function to be represented). – I mentioned here these works of *Chebyshev* about interpolation first of all because they are very valuable per se, and secondly because they are still not well known in Germany. Even Julius **[84]**
Bauschinger's report on interpolation, in issue n. 6 of the first volume of the Mathematische *Enzyklopädie* (I D3), whose reading I am recommending by the way, does not contain the deserved appreciation of Chebyshev's works. The presentation to the German mathematical audience of a summary report of these works would be welcome.

V. Functions of Two Variables

A quick review of the structure so far exposed of the lecture course shows that both in treating the variables x and the functions of one variable $f(x)$ we always had to stress the contrast between approximation and precision mathematics. The various practical applications of mathematics were then linked to approximation mathematics.

Geometry lies, in a manner of speaking, in the middle. Here we are using geometry mainly to facilitate the understanding of abstract considerations. The matter is as follows: Insofar as we treat curves that we can draw and imagine concretely we primarily explain relations of approximation mathematics; on the other hand, this geometric representation allows us to refer to the ideal object and thus to indirectly clarify its meaning, like – for instance – we clarified the essence of Weierstraß function by considering the successive approximating curves[89].

The matter is exactly the same when we move on to the *functions of two variables*. Here, too, we start with explanations that belong to the field of precision mathematics; my aim is to clarify certain fundamental relations – by constructing geometric figures – better than through the usual abstract representation.

Continuity

Let me first say something about the *domain of the variables* x, y. With the questions: "which is the more general domain?" or "when is this domain a continuum?" we would deeply enter the field of set theory. I will deal with this only later; for the moment, I only refer to the aforementioned report by *Arthur Rosenthal* in the *Enzyclopädie*.

For the sake of simplicity, suppose that our domain is a circle or a rectangle, whose sides are chosen parallel to the x- and y-axes (Fig. 42); and suppose our **[85]**
function is defined in this domain. What is meant is conceptually clear.

[89] After all, we could use all the areas of applications of mathematics for the same purpose for which we are using geometry here; but we do not want to enter too deeply into such speculations.

F. Klein, *Elementary Mathematics from a Higher Standpoint*,
DOI 10.1007/978-3-662-49439-4_6

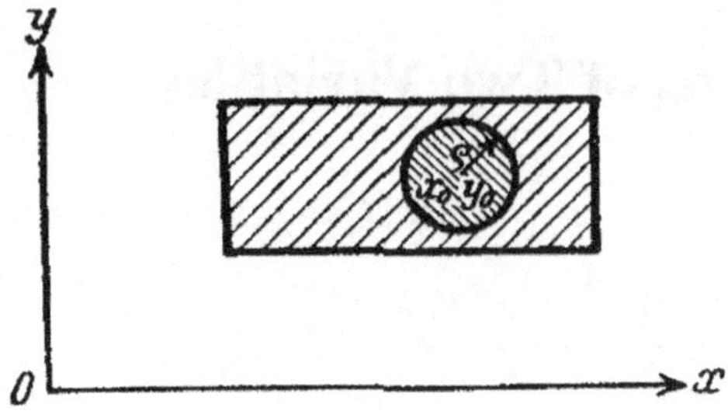

Figure 42

We only need at first to assume that to every pair of values $(x; y)$ of the domain corresponds a unique value z. But in any case it needs to be expressly indicated whether the boundary points have to belong to the domain or not; in the first case, we call the domain *closed*, in the other case *open*. Recently, the term "domain" is used in the sense of "closed domain", and the term "region"[90] is used for an "open domain".

When will I call *continuous* a function $f(x, y)$ at a point x_0, y_0? You will soon assent to the following definition: Let us consider the difference

$$|f(x, y) - f(x_0, y_0)|$$

and call a function $f(x, y)$ continuous at the point x_0, y_0 firstly if it is defined univocally at this point, and secondly if for any given positive δ, however small, one can find a ϱ different from zero such that, for every x, y for which $(x - x_0)^2 + (y - y_0)^2 < \varrho^2$ one obtains

$$|f(x, y) - f(x_0, y_0)| < \delta.$$

Obviously, we choose a circle of radius ϱ centred in x_0, y_0 only for the sake of simplicity; the important thing is that for any given δ, which may be as small as we wish, we can delimit a domain around x_0, y_0 within which remains $|f(x, y) - f(x_0, y_0)| < \delta$.

As you can see, this abstract definition of continuity does not yet present difficulties. But we soon find more complicated situations even when looking at quite simple analytic expressions, for instance the rational function

$$z = \frac{2xy}{x^2 + y^2}$$

This function is continuous at all points (x, y), with the exclusion of the point $(0, 0)$, where it is not defined because for $x = 0$ and $y = 0$ both numerator and denominator are equal to zero. The question is whether the function can be defined at $(0, 0)$ in such a manner as to become continuous therein.

[90] Translator's note: The two terms are today usually interchangeable.

If we introduce the polar coordinates

$$x = r\cos\varphi, y = r\sin\varphi$$

we obtain, for $r \neq 0$, a value that is independent from r:

$$z = \sin 2\varphi.$$

The most natural manner will be to extend the definition of the function so that it takes on the value $z = \sin 2\varphi$ also for $r = 0$. Its geometric representation is a regular surface of the 3rd order, which in mechanics bears the name "cylindroid". As is shown in Fig. 43 – taken from the work of *Robert Stawell Ball*: *A treatise on the theory of screws*, Cambridge 1900 – this surface is made of straight lines parallel to the plane xy, which cut the z-axis. **[86]**

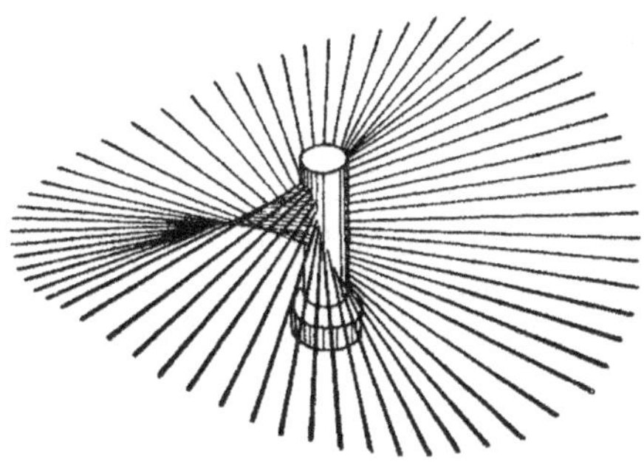

Figure 43

Did we obtain, thanks to our definition, that the function z is continuous at the point $x = 0, y = 0$? Apparently not, as it emerges from looking at the figure: the function is not single-valued at the point $(0, 0)$; it can take there any value from -1 to $+1$. But, undoubtedly, z is continuous along each path that leads to the origin. In fact, if a curve $y = g(x)$ passes trough the origin with the slope tg φ and a point (x, y) tends to $(0, 0)$ along this curve, so z tends to sin 2φ, that is, to a value that the function $f(x, y)$ really can take on for $x = 0$; $y = 0$.

If we had defined the function at the point $(0, 0)$ univocally, assigning it for instance the value 0, this would have fulfilled the first part of our definition of continuity, but we would not have guaranteed continuity along any path, and therefore not, as one easily recognizes, the second part of our definition of continuity. We will say:

Even for rational functions of two variables we encounter the phenomenon of continuo-ambiguity, which is at first excluded by our conceptual definition of function.

In addition to this first difficulty that can arise with rational functions of two variables, I want to explain a second one:

Suppose that a function $f(x, y)$ is *univocally* defined at the point x_0, y_0; moreover, it is continuous for any ray, determined by the azimuth Θ, along which we tend toward the point x_0, y_0 (Fig. 44). *Is then $f(x, y)$ continuous in x_0, y_0 in the sense of our previous general definition*?

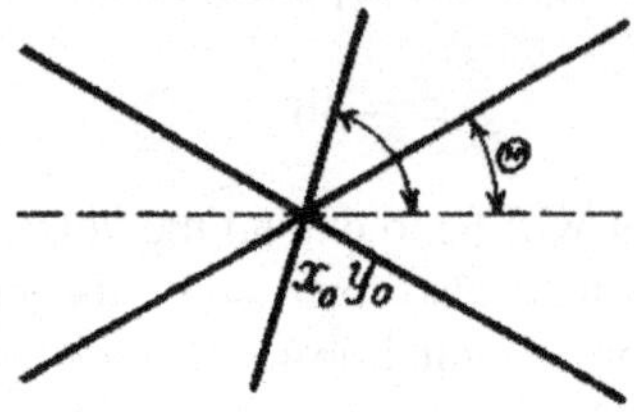

Figure 44

Expressed more in detail:

Suppose that for every Θ and for any positive δ, however small, we can assign a ϱ_Θ (the index shows the dependence from Θ) in such a way that for all $r < \varrho_\Theta$ we have

$$|f(x_0 + r\cos\Theta, y_0 + r\sin\Theta) - f(x_0, x_0)| < \delta.$$

Is then $f(x, y)$ per se continuous in the point x_0, y_0?

It turns out that this is not necessarily the case, so a function being continuous along every direction starting from x_0, y_0 does not necessarily need to be per se continuous.

[87] To explain this, I come back to what we have learned from the not uniform convergence of the Fourier series (also see the previous explanations about uniform continuity pp. [31]–[33]).

The non-uniform, that is the indefinitely slowed, convergence of the Fourier series at a place of discontinuity x_0 for a function that satisfies Dirichlet's conditions can be visualized with the help of a (step-)function. In fact we can determine – for each x singled out in a suitable neighbourhood of x_0 (for instance at its left side) – the minimal number n of terms that are necessary to obtain a determined degree of approximation, and then draw the quotient $\frac{1}{n}$ as the value of the ordinate corresponding to each x. The curve that we obtain in this way (Fig. 45), will come as close as desired to x_0 but it will have in x_0 an ordinate different from zero, as shown schematically in Fig. 45. Although the convergence is indefinitely slowed when approaching x_0, there is convergence at all points!

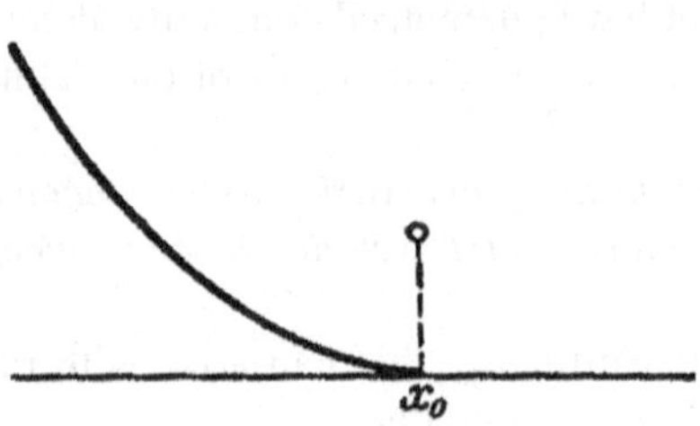

Figure 45

Something similar happens here with the *azimuthal* continuity[91].

We assume a positive δ chosen in a suitable way and fixed once and for all; we can further calculate for every ray departing from x_0, y_0 with azimuth Θ the upper bound of ϱ_Θ and apply this as a radius vector on the corresponding ray starting from x_0, y_0. So we obtain a curve (Fig. 46), which illustrates the degree of the azimuthal continuity for the different values of Θ. Now, although a ϱ_Θ different from zero exists on this curve for any Θ, it is nevertheless possible that ϱ_Θ falls below any fixed value ϱ in proximity of particular Θ-values. This happens when the curve – in proximity of certain azimuth values – gets arbitrarily close to the point x_0, y_0, then moving again, for this very azimuth, to a distance different form zero. I sum up: If we have only azimuthal continuity, then it is not excluded that ϱ_Θ, when approaching particular values of Θ, comes arbitrarily close to zero, even if, for the same values Θ, it takes on again a positive value.

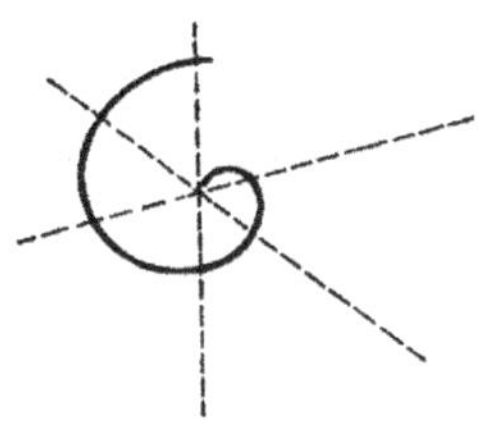

Figure 46

In such a case it is obviously not possible to describe a circle centred in x_0, y_0 with a fixed radius $\varrho > 0$ in such a way that for *all* points inside the circle the absolute value of the difference $f(x, y) - f(x_0, y_0)$ remains smaller than δ.

We can obtain an example of a function that has azimuthal continuity, but not continuity per se, in the following way. Suppose that Θ is the azimuth of the radius **[88]**
vector, which goes from the origin to the point (x, y). We can write $\Theta = \operatorname{arc}(x, y)$ for all value couples x, y which differ from the value couple $0, 0$. We define the following function:

$$\begin{aligned} F(x, y) &= \left(\pi^2 - \Theta^2\right) \sqrt[3]{1 : \Theta} && \text{for } \Theta \neq 0, \\ F(x, y) &= 0 && \text{for } \Theta = 0, \\ F(0, 0) &= 0. \end{aligned}$$

This function is single-valued for every x, y. Moreover, suppose that r is equal to $\left|\sqrt{x^2 + y^2}\right|$ and that p is any positive number. Then the function

$$z = f(x, y) = r^p F(x, y)$$

is only azimuthally continuous at the point $(0, 0)$.

That the function z is azimuthally continuous is immediately clear. In fact, $F(x, y)$ is constant along any radius vector, while r^p is continuous for every (x, y).

[91] Translator's note: Azimuthal continuity corresponds to continuity along a path; the function is continuous if it has the same limit computed along every path that approaches x_0, y_0 with the slope $\tan\Theta$, where Θ is the azimuth.

If we consider a circle centred in the origin, whose radius is just δ, then we have for all points internal to the circle:

$$|f(x,y)-f(0,0)|=|f(x,y)|<|\delta^p F(x,y)|,$$

and indeed $|f(x,y)|$ comes arbitrarily close to $|\delta^p F(x,y)|$ if (x,y) tends to the border of the circle. While the fixed δ may be a positive number as small as one wishes, we can nevertheless – by choosing arc (x,y) even smaller – succeed in obtaining $|F(x,y)|$ and therefore $|\delta^p F(x,y)|$ as large as we want.

Therefore azimuthal continuity is not the same as continuity per se. In order that azimuthal continuity becomes continuity per se we have to expressly require that one deals with uniform azimuthal continuity (that means exactly, it is possible to give a ϱ valid for *all* Θ and different from zero). This subtle question should thus be settled.

For the functions that are continuous in a *closed* domain we now have, analogously to the theorems for a function in one variable, the theorem of the existence of a *maximum and minimum value* and of *passing continuously through all intermediate values*, as you should make clear to you by yourself.

The obvious questions are now:

When is a continuous function $f(x,y)$ differentiable, indefinitely differentiable and can be expanded into a Taylor series?

First of all we have to deal with the partial derivatives of a function

$$\frac{\partial f}{\partial x}=p,\quad \frac{\partial f}{\partial y}=q.$$

Of course, they do not need to exist; rather we postulate their existence and ask for the higher derivatives, initially for the second derivatives:

$$\frac{\partial^2 f}{\partial x^2}=r,\quad \frac{\partial^2 f}{\partial x\partial y},\quad \frac{\partial^2 f}{\partial y\partial x},\quad \frac{\partial^2 f}{\partial y^2}=t.$$

[89] Now we must say:

If a continuous function $f(x,y)$ should have at any point not only the first derivative, but also the second and the higher order ones, then one not only needs to postulate their existence expressly, but one has in particular to investigate whether differentiation is independent from the order, that is whether, for instance, it results $\dfrac{\partial^2 f}{\partial x\partial y}=\dfrac{\partial^2 f}{\partial y\partial x}$.

I will explore the latter question later on. We now begin by assuming the existence of the derivative and also in general the commutability of the differentiations, and ask ourselves the question:

Can $f(x,y)$ then be expanded into the Taylor series

$$\begin{aligned} f(x,y)=f(x_0,y_0)&+\frac{(x-x_0)\,p_0+(y-y_0)\,q_0}{1!}\\ &+\frac{(x-x_0)^2\,r_0+2\,(x-x_0)\,(y-y_0)\,s_0+(y-y_0)^2\,t_0}{2!}\\ &+\cdots\cdots\cdots \end{aligned}$$

$$\left(s=\frac{\partial^2 f}{\partial x\partial y}=\frac{\partial^2 f}{\partial y\partial x}\right)?$$

According to our previous observations we have to give the following answer:

The existence of the derivatives and the convergence of the formally established series is absolutely not sufficient to have Taylor's expansion, additionally one has to require *Pringsheim's conditions*, which means that the remainder has to come arbitrarily close to zero as n increases. Only if all this holds, can we call $f(x, y)$ – for the moment in the domain of convergence of the series – an analytic function of the two variables x, y.

And we might add: *A great number of requirements have to be satisfied at the same time so as to have an analytic function* $f(x, y)$.

Now we turn briefly to the applications and ask:

What about the functions in two ore more variables used in mechanics and physics?

The usual mode of expression in the applications is such that every function is seen as analytic and one considers oneself entitled to dismiss series terms higher than those of the first, second or third order "because this yields a sufficient approximation".

We have to expose our criticism against this.

Surely, this mode of expression does not take into consideration the exact concept of analytic function described above. Rather, one is acting from the beginning within the field of approximation mathematics and simply assumes that the relations found can be represented with sufficient approximation by polynomials of the first, second, or third degree. So, considering the terminology used, the main weight is posed on the second part of the assertion: *The assumption is that func-* [90]
tions occurring in the applications can be represented with a sufficient accuracy by polynomials of the first, second, or third degree. The critical conception that I am representing here does not concede that the occurring functions are "in reality" analytic; we have no means to affirm this.

Interchangeability of the Order of Differentiation. Practical Example of a Function for Which $\frac{\partial^2 f}{\partial x\,\partial y} \neq \frac{\partial^2 f}{\partial y\,\partial x}$

The question is: Is it true that, in the applications, the only relations that occur are those for which the mentioned approximation is allowed; in particular (given the importance that derivatives generally have in the applications) is it always $\frac{\partial^2 f}{\partial x \partial y} = \frac{\partial^2 f}{\partial y \partial x}$ in the practice? Or is the entire assumption only the common one, suggested by habit and not by actual facts?

We must say: *there exist in practice enough examples where we are not allowed to put* $\frac{\partial^2 f}{\partial x \partial y} = \frac{\partial^2 f}{\partial y \partial x}$.

To clarify this, we first ask:

Under which conditions does one prove, within precision mathematics, that $\frac{\partial^2 f}{\partial x \partial y} = \frac{\partial^2 f}{\partial y \partial x}$ holds? Later on we will construct examples, in which these conditions are not satisfied, looking among those surfaces that practice itself presents us with.

The definition of $\frac{\partial^2 f}{\partial y \partial x}$ is the following:

First we have

$$\frac{\partial f}{\partial x} = \lim_{k \to 0} \frac{f(x, y+k) - f(x, y)}{k} = q(x, y)$$

and then

$$\frac{\partial q}{\partial x} = \frac{\partial^2 f}{\partial y \partial x} = \lim_{h \to 0} \frac{q(x+h, y) - q(x, y)}{h}.$$

Unifying the two equations we have

$$s' = \frac{\partial^2 f}{\partial y \partial x} = \lim_{h \to 0} \left(\lim_{k \to 0} \frac{f(x+h, y+k) - f(x+h, y) - f(x, y+k) + f(x, y)}{hk} \right)$$

therefore $\frac{\partial^2 f}{\partial y \partial x}$ results from the fraction on the right hand side if first k and then h tend to zero.

Analogously we have:

$$s = \frac{\partial^2 f}{\partial x \partial y} = \lim_{k \to 0} \left(\lim_{h \to 0} \frac{f(x+h, y+k) - f(x+h, y) - f(x, y+k) + f(x, y)}{hk} \right)$$

so that we obtain the other derivative $\frac{\partial^2 f}{\partial x \partial y}$ when we perform the two limit passages in the inverse order, that is we let first h and then k tend to zero.

[91] *The question of equality of the two derivatives is hence identical to that of whether the two limit passages are interchangeable or not.*

Now, using the mean value theorem, we will elaborate the usual case, in which in fact $s = s'$ holds.

We have

$$p(x, y) = \lim_{h \to 0} \frac{f(x+h, y) - f(x, y)}{h}$$

Under the condition that p exists in (x, y) and in a neighbourhood U of (x, y) (namely, p is univocal and finite in U), the mean value theorem leads, for a suitable choice of h, to the relation

$$f(x+h, y) - f(x, y) = h \cdot p(x + \vartheta h, y),$$

where $0 < \vartheta < 1$.

Now we consider p in U as differentiable at y; with a suitable choice of k and applying again the mean value theorem, we have

$$p(x + \vartheta h, y + k) - p(x + \vartheta h, y) = k \cdot s(x + \vartheta h, y + \eta k),$$

where $0 < \eta < 1$, and, summarizing

$$f(x+h, y+k) - f(x, y+k) - f(x+h, y) + f(x, y) = h \cdot k \cdot s(x + \vartheta h, y + \eta k).$$

We divide both members of the equation by k and write:

$$\frac{f(x+h,y+k)-f(x+h,y)}{k}-\frac{f(x,y+k)-f(x,y)}{k}=h{\cdot}s(x+\vartheta h,y+\eta k).$$

If we now assume the existence of q in (x,y) and U, in the limit passage $k\to 0$ the left hand side of the equation becomes

$$q(x+h,y)-q(x,y).$$

But for the right hand side it holds – with the further assumption that, for constant values of x, s is continuous for y in the neighbourhood U:

$$\lim_{k\to 0} s\,(x+\vartheta h,y+\eta k)=s\,(x+\vartheta h,y)\,.$$

Consequently

$$\frac{q\,(x+h,y)-q\,(x,y)}{h}=s\,(x+\vartheta h,y)\,.$$

Finally, we also assume that s is continuous at (x,y) with respect to x. We then let h tend to 0 and obtain

$$\lim_{h\to 0}\frac{q\,(x+h,y)-q\,(x,y)}{h}=s\,(x,y)\,.$$

The limit on the left side is nothing else than $s'(x,y)$. Therefore: $s'(x,y)$ exists and is equal to $s(x,y)$.

In our proof we did not only have to assume the existence of p,q,s at the point (x,y) and in a neighbourhood U of (x,y), but also certain continuity conditions for **[92]** s. If we drop the one or the other of these assumptions we can no longer expect that s' – if it exists – is equal to s[92].

Now we want to show that among the surfaces that we observe every day[93] there are in fact examples in which $s'\neq s$.

I take this example from *architecture*.

Suppose that two cylindrical vaults with congruent semi-circular cross sections that cross each other forming two "intersecting vaults" (Fig. 47). We choose the apex as origin of the coordinates, the z-axis directed vertically upwards, and we now want to calculate the values s and s' for the origin.

[92] [The question of the interchangeability of the order of differentiation in the case described above has been treated thoroughly for the first time by *Hermann A. Schwarz* in his paper: Über ein vollständiges System von einander unabhängiger Voraussetzungen zum Beweise des Satzes: $\frac{\partial}{\partial y}\left(\frac{\partial f(x,y)}{\partial x}\right)=\frac{\partial}{\partial x}\left(\frac{\partial f(x,y)}{\partial y}\right)$ (*Hermann A. Schwarz, Gesammelte mathematische Abhandlungen*, Vol. 2, Berlin: Springer, 1890, pp. 275–284). For functions of two or more variables and partial derivatives of order higher than 2, the study has been conducted by *Ludwig Neder* (Über Funktionen von reellen Argumenten, *Mathematische Zeitschrift*, Vol. 24 (1926), pp. 759–784)].

[93] Or, more precisely (as we never consider pure surfaces): Among the idealized surfaces, that everyday observation suggests the study of.

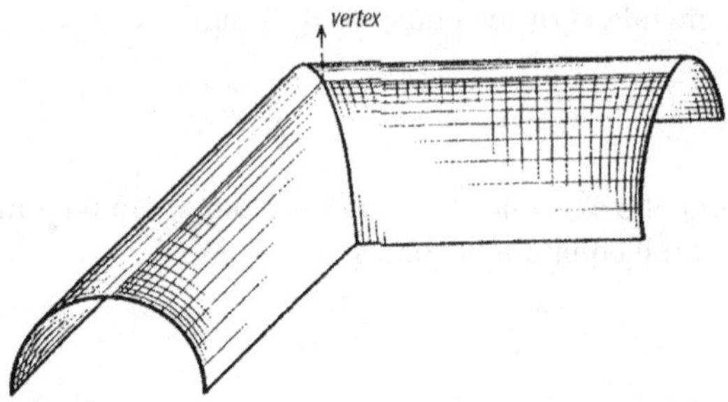

Figure 47

To obtain a usable result that is not hidden by the symmetry of the vault, we imagine the whole vault oriented by an angle α, with $0 < \alpha < 45°$ with respect to the coordinate system xy; so that the projection onto the xy-plane has the position shown in Fig. 48.

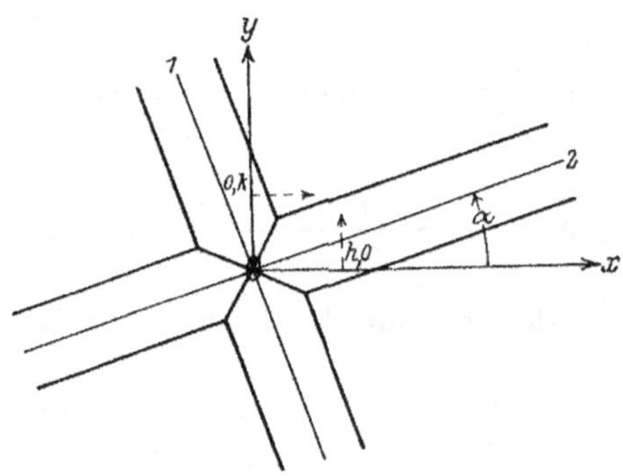

Figure 48

For each point of the vault we calculate p and q as limit values of difference quotients. In the apex we will have in any case $p = 0, q = 0$, because along the central line of each vault the tangent plane is horizontal; so $p(0,0) = q(0,0) = 0$.
[93] What are now $s(0,0)$ and $s'(0,0)$? We define them as limits as follows:

$$s = \lim_{k \to 0} \frac{p\,(0,k) - p\,(0,0)}{k}, \quad s' = \lim_{h \to 0} \frac{q\,(h,0) - q\,(0,0)}{h}$$

For the sake of simplicity, we choose h and k positive, so point $0, k$ lies on the vault 1; with p we measure the increase and decrease of vault 1 in the direction of x. Consequently, $p(0,k)$ is negative. Instead, the point $h, 0$ lies on vault 2; with q we measure the increase and decrease of the vault in the direction of y. Consequently, $q(h,0)$ is positive. From this qualitative consideration follows:

s originates with a limit passage from a negative difference quotient, s' originates with a limit passage from a positive difference quotient. Thus, unless s and s' happen to become both zero, s cannot be equal to s'.

In order to follow the calculation of the passage to the limit, we need to introduce some formulas.

For the original position, in which the central lines of the vaults are parallel to the axes x, y, we have the formulas[94]:

$$z = C_0 - C_1 y^2 + \cdots \text{ for vault 2,}$$
$$z = C_0 - C_1 x^2 + \cdots \text{ for vault 1;}$$

the linear terms drop due to $p(0,0) = 0$ and $q(0,0) = 0$; the terms of order higher then two are omitted for simplicity, because they would have no influence in the result to be derived.

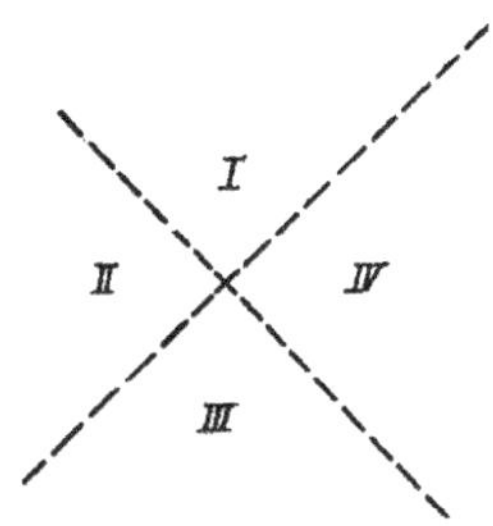

Figure 49

Let us now consider the position shown in Fig. 48 and denote (as indicated in Fig. 49[95]) the different parts of the entire vault by I, ..., IV, then we have to put for I and III (vault 1)

$$z = C_0 - C_1(x \cos\alpha + y \sin\alpha)^2 - \cdots,$$

for II and IV (vault 2)

$$z = C_0 - C_1(-x \sin\alpha + y \cos\alpha)^2 - \cdots.$$

We obtain[96]
for I and III:

$$p = -2C_1 x \cos^2\alpha - 2C_1 y \sin\alpha \cos\alpha - \cdots$$
$$q = -2C_1 x \sin\alpha \cos\alpha - 2C_1 y \sin^2\alpha - \cdots,$$

for II and IV:

$$p = -2C_1 x \sin^2\alpha + 2C_1 y \sin\alpha \cos\alpha - \cdots$$
$$q = +2C_1 x \sin\alpha \cos\alpha - 2C_1 y \cos^2\alpha + \cdots.$$

[94] Translator's note: The expansion into power series, stopped at the second term, yields in this case an approximation with a parabolic cylinder.

[95] Translator's note: The figure is explained with more details in the version of 1902. The main diagonal corresponds to the new position of vault 2 (II and IV in the old position). Similarly the old position of vault 1 is given by I and III.

[96] Translator's note: Using the partial derivatives of z.

[94] According to these formulas, in the apex of the intersecting vaults in fact we have, for both vaults, $p = 0, q = 0$. But for the rest, along the intersection curves of the vaults, a discontinuity occurs in the values p and q, the two vaults run into each other at an edge.

Now we calculate s in the origin. From $p(0, k)$ in I we obtain, leaving out the higher terms of the series expansion:

$$s = \lim_{k \to 0} \frac{-2C_1 k \sin \alpha \cos \alpha}{k} = -C_1 \sin 2\alpha.$$

Instead, for s' we obtain in IV, omitting the higher terms,

$$s' = \lim_{h \to 0} \frac{+2C_1 h \sin \alpha \cos \alpha}{h} = +C_1 \sin 2\alpha.$$

Thus, in general, we have $s \neq s'$, and it is $s = s'$ only when the coordinate axes are parallel to the centre lines.

If we calculate the values of s at points that are in the neighbourhood of $(0, 0)$, in the parts II and IV of the vaults, and the value s' for the analogous points in parts I and III, we obtain – excluding terms of higher order –

$$s = +C_1 \sin 2\alpha, \quad s' = -C_1 \sin 2\alpha.$$

So one can see that both $s(x, y)$ and $s'(x, y)$ are discontinuous at the point $(0, 0)$.

As you can see, this example is neither abstract nor difficult, but rather commonplace. We can generalize it, for instance, so as to obtain a surface of the form of an umbrella-shaped roof. If the umbrella is held upright, but not symmetrically with respect to the xz- and xy-planes, the discussed case $s \neq s'$ occurs, at the top – above our heads. So we can perhaps conclude in the following way:

The surface shapes for which s and s' are different in the sense of the example described here are basically commonplace, and it is only due to our education – lacking in independent tinkering and precise observation of concrete relations – that we do not think about this (and always accept without criticism the approaches that we find in books).

This is all I will say about functions of two variables regarding precision mathematics. Now I am switching to some observations about them from the point of view of approximation mathematics, similarly to what I did before, when – after the treatment of the functions $f(x)$ in precision mathematics – a chapter on interpolation and approximation followed.

Approximate Representation of Functions of the Spherical Surface by Means of Series of Spherical Harmonics[97]

How can one approximate a function $f(x, y)$?

The most obvious thing is to approximate $f(x, y)$ *through homogeneous polynomials in x, y of increasing degree*:

$$z = f_0 + f_1 + f_2 + \cdots,$$

what is often used in mechanics and physics, as we mentioned. **[95]**

Another possibility is to try to approximate the function by *Laplacian spherical harmonics*.

I omit the first point and intend to only say *something about Laplace's spherical harmonics*, enough to make their essence and their descriptive power understandable.

The issue is hence a generalization of the trigonometric series. To this purpose, while now denoting the argument by ω and not by x as before, we need to formally introduce the series for $f(\omega)$

$$f(\omega) = \frac{a_0}{2} + a_1 \cos\omega + \cdots + a_n \cos n\omega + \cdots,$$
$$+ b_1 \sin\omega + \cdots + b_n \sin n\omega + \cdots,$$

in a slightly different way than hitherto.

1. We interpret ω as central angle and imagine the values z of a function $f(\omega)$ of period 2π assigned along the periphery of a circle; the radius can be thought of as unitary (Fig. 50).

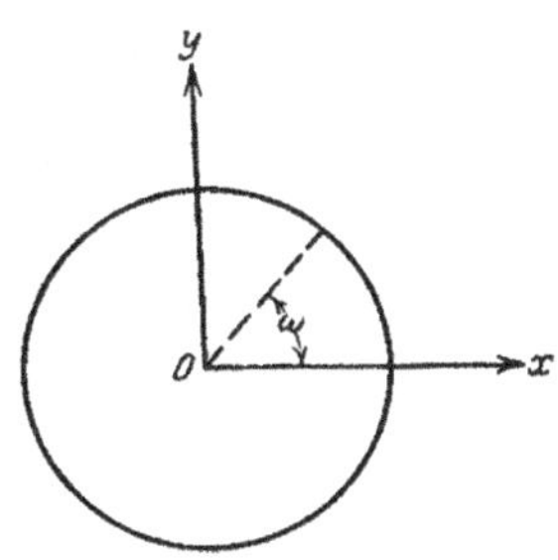

Figure 50

[97] Translator's note: Klein uses the term spherical functions.

2. The function $f(\omega)$ is now approximated by pairs of single terms $a_n \cos n\omega + b_n \sin n\omega$, whose mathematical properties we express in a somewhat different manner introducing orthogonal coordinates.

It is $\cos\omega = x, \sin\omega = y$, and therefore

$$\cos\omega + i \sin\omega = x + iy,$$

and, using de Moivre,

$$\cos n\omega + i \sin n\omega = (x + iy)^n, \cos n\omega - i \sin n\omega = (x - iy)^n.$$

In a well known manner one can thus express

3. $\cos n\omega$ and $\sin n\omega$ in the form of polynomials of the n-th degree in x, y. The same holds for $a_n \cos n\omega + b_n \sin n\omega$. All these polynomials F satisfy the simple differential equation

$$\Delta F = \frac{\partial^2 F}{\partial x^2} + \frac{\partial^2 F}{\partial y^2} = 0,$$

because both $(x + iy)^n$ and $(x - iy)^n$ satisfy it. The most important thing now is that this differential equation characterizes our polynomials.

In this sense:

4. If I search an homogeneous polynomial of the n-th degree that has to satisfy the equation $\Delta F = 0$ then I naturally think of a linear combination $a \cos n\omega + b \sin n\omega$.

This follows from a simple calculation: A homogeneous polynomial F of degree
[96] n has $(n + 1)$ coefficients. In the corresponding ΔF we have a homogeneous polynomial of degree $(n - 2)$ with $(n - 1)$ coefficients, which should be identically zero.

This way the $(n + 1)$ coefficients have to satisfy $(n - 1)$ conditions; two coefficients remain arbitrary (a and b in our expression).

If we denote the polynomials of degree 0 to n mentioned by $F_0, F_1, \ldots, F_n$, we obtain the following result:

The usual trigonometric series for $f(\omega)$ can be designated as a series expansion that proceeds by homogeneous polynomials in x, y of increasing degree, where each polynomial F_υ satisfies the characterizing equation $\frac{\partial^2 F_\upsilon}{\partial x^2} + \frac{\partial^2 F_\upsilon}{\partial y^2} = 0$. Thereafter we put $x = \cos\omega$, $y = \sin\omega$.

Here we are not yet facing the question as to whether we have to do with an infinite or a finite series, but only the formal expression of the terms is envisaged. Furthermore, the art lies in the suitable calculation of the constants a_n, b_n, which are at the moment still arbitrary.

By describing the usual trigonometric series in this manner they become capable of being generalised for several variables. We will increase now by only one dimension.

We imagine a sphere of radius 1 and a point on it given by its coordinates x, y, z. If we introduce spherical coordinates φ, ϑ, where φ is the geographical longitude, going from 0 to 2π, and ϑ the polar distance, from 0 to π, we obtain the following formulas (Fig. 51):

$$x = \sin\vartheta\cos\varphi,$$
$$y = \sin\vartheta\sin\varphi,$$
$$z = \cos\vartheta.$$

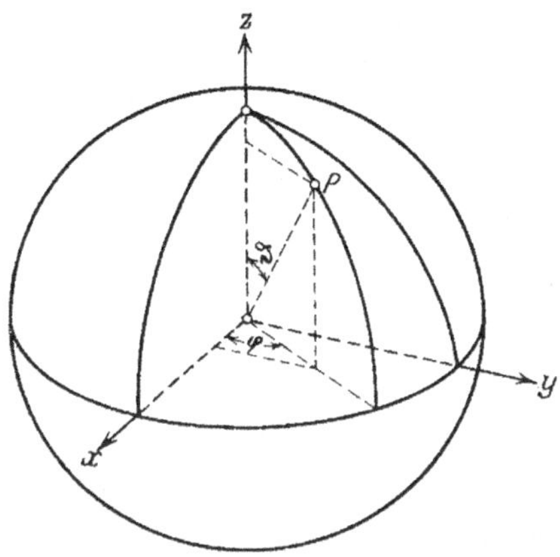

Figure 51

Let us now consider a function $f(\vartheta, \varphi)$ on the sphere. We want to see if we can use it to represent an analogous series of homogeneous polynomials of degree $\nu(\nu = 0, 1, 2, \ldots n)$ in x, y, z, which now satisfy the equation

$$\Delta F_\nu = \frac{\partial^2 F_\nu}{\partial x^2} + \frac{\partial^2 F_\nu}{\partial y^2} + \frac{\partial^2 F_\nu}{\partial z^2} = 0.$$

Consequently

$$f\,(\vartheta, \varphi) = F_0 + F_1 + \cdots + F_n \cdots.$$

It constitutes a special issue whether the series converges – in the case that it is infinite – or whether the finite series somehow suitably approximates our function $f(\vartheta, \varphi)$ and we will not tackle it for the moment. [97]

The question is only to elaborate the formal rule that can be understood as a natural generalization of what we have done before.

We call a single polynomial F a *Laplace spherical harmonics*, so that we obtain the following definition:

By a (Laplace) spherical harmonics we firstly mean a homogeneous polynomial of the n-th *degree in* x, y, z *that satisfies the differential equation* $\Delta F = 0$.

Later on we will interpret a spherical harmonic, in a narrower sense, as the expression obtained by substituting to x, y, z their expression in φ, ϑ.

We now consider the number of constants in F_n that will still remain undetermined. An arbitrary $F_n(x, y, z)$ has $\frac{(n+1)(n+2)}{2}$ constants (this can be proved by induction from the fact that F_1 has three constants and F_2 six). When one forms ΔF, one obtains a polynomial of degree $(n-2)$ with $\frac{(n-1)n}{2}$ constants. To ensure that it vanishes identically the coefficients will have to satisfy $\frac{(n-1)n}{2}$ conditions; therefore

$$\frac{(n+1)(n+2)-(n-1)n}{2} = \frac{4n+2}{2} = 2n+1$$

constants will remain arbitrary, so that we obtain the theorem:

The more general spherical function of degree n still contains $2n+1$ *undetermined parameters*; actually these parameters are linear, because from $\Delta F = 0$ we only obtain linear conditions for the coefficients of F.

We verify this for $n = 2$. The most general polynomial of the second degree is

$$F_2 = a_{11}x^2 + a_{22}y^2 + a_{33}z^2 + 2a_{23}yz + 2a_{31}zx + 2a_{12}xy.$$

From $\Delta F_2 = 0$ it results

$$2(a_{11} + a_{22} + a_{33}) = 0.$$

Therefore, 6 coefficients are subjected to linear conditions, and 5 coefficients remain still arbitrary, as it should be.

We actually want to construct the polynomials in the four lowest cases. This is usually done by referring to the representation in spherical coordinates and giving by means of a particular table (see the table on p. [98]) $2n+1$ special spherical functions for each $n = 0, 1, 2, 3, 4, \ldots$; the general spherical function of the corresponding degree can thus be obtained as the sum of these special spherical functions still having arbitrary constant factors.

[98] I observe that the first vertical line of the table needs to be calculated[98], while the other lines can be easily obtained by differentiating in $\cos\vartheta$ the first terms – omitting certain coefficients – and multiplying them by the given factors.

The most general spherical function F_4 is obtained linearly from the nine written special spherical functions F_4, with arbitrary coefficients.

To perform a precise task, we want to approximate a given function $f(\vartheta, \varphi)$ as well as possible by means of the first four spherical functions, according to the least square method. That is, we put

$$f(\vartheta, \varphi) = F_0 + F_1 + F_2 + F_3 + F_4 + \text{remainder}$$

and ask for the minimum of the sum of the squares of the errors, that is of the integral calculated over the surface of the sphere with the element do

$$\int (f - F_0 - F_1 - \cdots - F_4)^2 \cdot do.$$

[98] The verification of the polynomials indicated is of course very simple. For instance, instead of

$$F_3 = 5\cos^3\vartheta - 3\cos\vartheta$$

going back to orthogonal coordinates one has to put

$$F_3 = 5z^3 - 3z(x^2 + y^2 + z^2)$$

and to verify that ΔF_3 vanishes.

Spherical functions up to the fourth degree

F_0	1				
F_1	$\cos\vartheta$	$\left.\begin{matrix}\sin\vartheta\cos\varphi\\ \sin\vartheta\sin\varphi\end{matrix}\right\} 1$			
F_2	$3\cos^2\vartheta - 1$	$\left.\begin{matrix}\sin\vartheta\cos\varphi\\ \sin\vartheta\sin\varphi\end{matrix}\right\} \cos\vartheta$	$\left.\begin{matrix}\sin^2\vartheta\cos 2\varphi\\ \sin^2\vartheta\sin 2\varphi\end{matrix}\right\} 1$		
F_3	$5\cos^3\vartheta - 3\cos\vartheta$	$\left.\begin{matrix}\sin\vartheta\cos\varphi\\ \sin\vartheta\sin\varphi\end{matrix}\right\} (5\cos^2\vartheta - 1)$	$\left.\begin{matrix}\sin^2\vartheta\cos 2\varphi\\ \sin^2\vartheta\sin 2\varphi\end{matrix}\right\} \cos\vartheta$	$\left.\begin{matrix}\sin^3\vartheta\cos 3\varphi\\ \sin^3\vartheta\sin 3\varphi\end{matrix}\right\} 1$	
F_4	$35\cos^4\vartheta - 30\cos^2\vartheta + 3$	$\left.\begin{matrix}\sin\vartheta\cos\varphi\\ \sin\vartheta\sin\varphi\end{matrix}\right\} (7\cos^3\vartheta - 3\cos\vartheta)$	$\left.\begin{matrix}\sin^2\vartheta\cos 2\varphi\\ \sin^2\vartheta\sin 2\varphi\end{matrix}\right\} (7\cos^2\vartheta - 1)$	$\left.\begin{matrix}\sin^3\vartheta\cos 3\varphi\\ \sin^3\vartheta\sin 3\varphi\end{matrix}\right\} \cos\vartheta$	$\left.\begin{matrix}\sin^4\vartheta\cos 4\varphi\\ \sin^4\vartheta\sin 4\varphi\end{matrix}\right\} 1$

So we do not take into consideration the case, mostly the sole case treated in textbooks, of the expansion of a function into an infinite series of spherical harmonics, and we only deal with the approximation through a finite series, which is the only one interesting in practice. Our task is to calculate the 25 unknown coefficients that appear in the expression $F_0 + F_1 + \cdots + F_4$ in such a way as to satisfy the minimum requirements.

[99] So it seems that we have to solve 25 linear equations of the most general kind, with 25 unknowns. But the situation is not so bad, because each particular equation is already solved with respect to just one unknown, as we will see soon.

To perform the calculation easily we use the summation sign and write for the spherical functions listed in the first vertical row of the table, in the usual manner, $P_n(\cos\vartheta)$. The others are then

$$\left.\begin{array}{l}\sin^\nu\vartheta\cos\nu\varphi\\ \sin^\nu\vartheta\sin\nu\varphi\end{array}\right\} P_n^{(\nu)}(\cos\vartheta),$$

where the index ν in P_n means the ν-th derivative in $\cos\vartheta$ (omitting the occurring numerical coefficients). So the whole series becomes

$$f(\vartheta,\varphi) = \sum_{n=0}^{4}\sum_{\nu=0}^{n}\left[(a_{n,\nu}\sin^\nu\vartheta\cos\nu\vartheta + b_{n,\nu}\sin^\nu\vartheta\sin\nu\vartheta)\, P_n^{(\nu)}(\cos\vartheta)\right],$$

for which I write shortly

$$\sum\sum(a_{n,\nu}\Phi_{n,\nu} + b_{n,\nu}\Psi_{n,\nu}).$$

If one first searches the sum of the squares of the errors,

$$\int\left\{f(\vartheta,\varphi) - \left(\sum\sum(a_{n,\nu}\Phi_{n,\nu} + b_{n,\nu}\Psi_{n,\nu})\right)\right\}^2 do,$$

one obtains a remarkable simplification for the terms quadratic in a and b, due to the so-called *property of orthogonality of spherical functions.*

In fact, if F' and F'' denote two different spherical functions of our special 25, then it is $\int F'F''do = 0$, or, in particular

$$\int\Phi'\Phi''do = 0,\quad \int\Phi'\Psi''do = \int\Phi''\Psi'do = 0,\quad \int\Psi'\Psi''do = 0^{99}$$

99 [Also see the analogous behaviour of the integrals in the trigonometric series: $\int_0^{2\pi}\cos\mu x\cos\nu x dx$, $\int_0^{2\pi}\sin\mu x\sin\nu x dx$ for $\mu\neq\nu$ and of the integral $\int_0^{2\pi}\sin\mu x\cos\nu x dx$ both for $\mu=\nu$ and $\mu\neq\nu$.]

This way the function Ω (namely the sum of the squares of the errors that has to be minimized through a suitable choice of the coefficients a, b) becomes much simpler than one might have assumed. It is denoted by

$$\Omega = \int f^2 do - 2\sum \int f \cdot F \cdot do + \sum \int F^2 do,$$

where the sums are to be taken over all the special spherical functions given above. **[100]**
Expressed in Φ and Ψ the function is

$$\Omega = \int f^2 do - 2\sum\sum \left[a_{n,\nu} \int f\Phi_{n,\nu} do + b_{n,\nu} \int f\Psi_{n,\nu} do \right]$$
$$+ \sum\sum \left[a_{n,\nu}^2 \int f\Phi_{n,\nu}^2 do + b_{n,\nu}^2 \int f\Psi_{n,\nu}^2 do \right].$$

To obtain its minimum, we put the partial differential equations in $a_{n,\nu}$ and $b_{n,\nu}$ equal to zero. We get

$$\frac{1}{2}\frac{\partial\Omega}{\partial a_{n,\nu}} = -\int f\Phi_{n,\nu} do + a_{n,\nu} \int \Phi_{n,\nu}^2 do = 0,$$
$$\frac{1}{2}\frac{\partial\Omega}{\partial b_{n,\nu}} = -\int f\Psi_{n,\nu} do + b_{n,\nu} \int \Psi_{n,\nu}^2 do = 0$$

Here the integrals have always to be extended over the whole surface of the sphere. As one remarks, each of the 25 equations thus obtained contains just one unknown; we easily get

$$a_{n,\nu} = \frac{\int f\Phi_{n,\nu} do}{\int \Phi_{n,\nu}^2 do}, \qquad b_{n,\nu} = \frac{\int f\Psi_{n,\nu} do}{\int \Psi_{n,\nu}^2 do},$$

Thus the problem is solved. We add the following general remark:

The beautiful aspect of this manner of proceeding is that (exactly as in Fourier's series) the value of the particular $a_{n,\nu}$ or $b_{n,\nu}$ – as it results from the minimum requirement for the sum of the squares of the errors – does not depend on the other pairs of indexes n, ν which one might use additionally for the approximation of f, so that, once a coefficient $a_{n,\nu}$ has been calculated it also remains valid when one wants later on to arbitrarily extend the series. So, if – for an approximation via spherical harmonics up to the fourth degree – we have calculated the 25 coefficients, these remain unvaried if we go on to the spherical functions of fifth degree; we only have to add 11 new coefficients.

All this concerned a purely formal way of constructing the series. For an *insight into the nature of the spherical harmonics, that for the moment has been defined purely formally*, we once again start from the Fourier series.

Distribution of the Values of a Spherical Function over the Sphere

The Fourier series is built on the terms $\cos n\varphi, \sin n\varphi$; it is easy to see how these behave on the periphery of the circle, in particular where the zeros are. The number of zeros is $2n$ and increases with n. At the same time, the zeros are equally distributed over the periphery (Fig. 52).

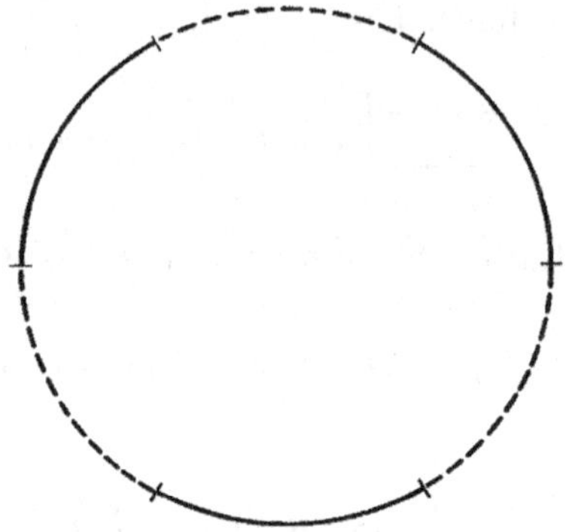

Figure 52

The functions P_n $(\cos\vartheta)$, that is $P_n(z)$, at first behave in a similar manner.

We consider a sphere and search the zeros of the function $P_n(z)$. Without giv-
[101] ing a proof here – as only a first overview on spherical harmonics is intended –
I maintain that:

$P_n(z)$ has n real zeros between $z = -1$ and $z = +1$, which lie symmetrical with respect to $z = 0$, that is: On the sphere there are n real parallel circles that lie symmetrically to the equator and on which $P_n(z) = 0$.

Therefore, such a spherical harmonics $P_n(z)$ divides the sphere in $(n + 1)$ zones in which $P_n(z)$ is alternately positive and negative. Therefore, the $P_n(z)$ is called – according to the English practice – "*zonal spherical function*" (Fig. 53).

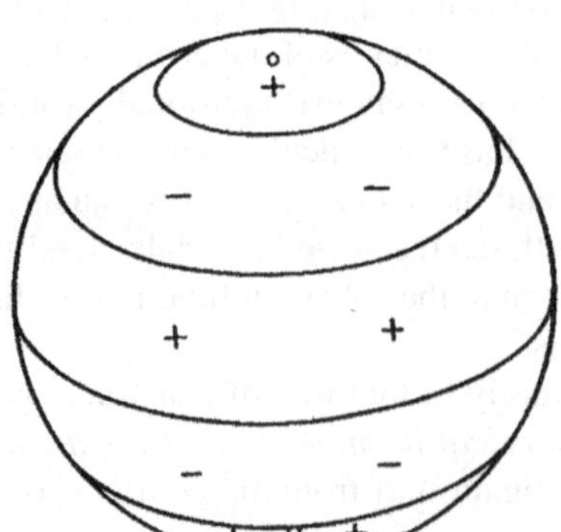

Figure 53

Besides the spherical harmonics P_n $(\cos\vartheta)$ we had put other functions with the factors $\left.\begin{matrix}\cos\nu\varphi\\ \sin\nu\varphi\end{matrix}\right\}\sin^\nu\vartheta$.

Let us right away consider the highest case $\nu = n$; there it results $P_n^{(n)}(\cos\vartheta) = 1$ and we have to deal with the spherical functions:

$$\left.\begin{matrix}\cos n\varphi\cdot\sin^n\vartheta\\ \sin n\varphi\cdot\sin^n\vartheta\end{matrix}\right\}1$$

If we ask for the zeros here, the relevant element is the first factor (the second one yields the poles of the sphere in each case). We obtain that:

$$\text{The spherical functions}\quad\left.\begin{matrix}\cos n\varphi\\ \sin n\varphi\end{matrix}\right\}\sin^n\vartheta$$

divide the sphere into $2n$ equiangular sectors (spherical bi-angles) delimited by meridians, for which the spherical functions are alternately positive and negative. The spherical functions considered are therefore called "*sectorial spherical harmonics*" (Fig. 54).

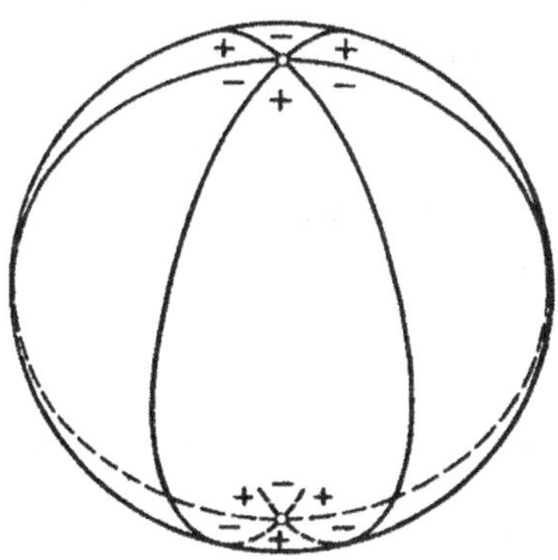

Figure 54

Now, for the intermediate cases. **[102]**

$$\left.\begin{matrix}\cos\nu\varphi\sin^\nu\vartheta\\ \sin\nu\varphi\sin^\nu\vartheta\end{matrix}\right\}P_n^{(\nu)}(\cos\vartheta)\,.$$

To the first factor, when it is put equal to zero, correspond ν meridians, which split the equator in arcs of equal length, to the second factor correspond $n-\nu$ parallel circles which are symmetric with respect to the equator (which is not difficult to recognise), so that the spherical surface – with the exception of the areas of the poles, that yield triangles – is split into particular quadrilaterals; this is why these spherical functions are occasionally called "*tesseral spherical harmonics*" (Fig. 55).

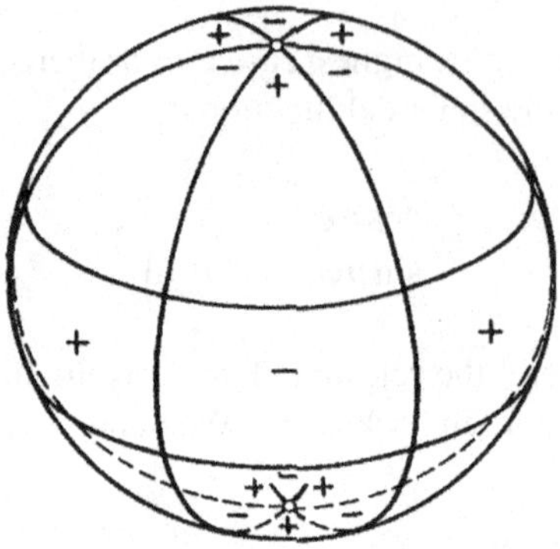

Figure 55

Analogously to $\cos n\varphi, \sin n\varphi$, which can be called *oscillating functions* along the periphery of the circle, also our spherical harmonics can have the same designation with *respect to the sphere*, in that we can clearly see how the zones with positive and negative function values are alternated. It is of course absolutely interesting to investigate this up to its numerical details, in particular to observe how the positive and negative zones of the different spherical functions overlap.

Estimate of the Error Truncating the Series of Spherical Harmonics

The series that we are studying is built out of such oscillating functions. *The question is to what extent the function $f(\vartheta, \varphi)$ is approximated at particular points*, that is, whether we can estimate the error.

On this question, I only report:

The error, that is the difference between the function to be represented and the finite sum, can be indicated at every point in a closed form by means of an integral extended over the sphere, again in complete analogy to what we know from the Fourier series.

The corresponding estimation is found in textbooks, however, the focus there is often put one-sidedly on the fact that, for indefinitely increasing n – in the case that $f(x, y)$ is a "reasonable" function – the remainder converges towards zero, while in practice it is important to estimate the remainder for a finite n numerically[100].

[100] [The infinite series that arise upon expanding the series into spherical functions, show in many respects a behaviour similar to that of the infinite Fourier series. See *Leopold Fejér*, Über die Laplacesche Reihe, *Mathematische Annalen*, Vol. 67 (1909), pp. 76–109, and the two papers of *Hermann Weyl* on Gibbs' phenomenon in the theory of the spherical functions, Vol. XXIX (1910),

How far one needs to go with n, that is, how many terms of the expansion of the spherical functions are eventually necessary to obtain a precision sufficient for each practical case, can in no way be deduced from our theoretical development and can be decided only from case to case. **[103]**

As a famous example I mention the publication of *Carl Friedrich Gauss* of the year 1839: *Allgemeine Theorie des Erdmagnetismus* (outcomes of the observations of the magnetic Association in the year 1838 = Werke, Vol. 5, pp. 121–193). There the potential V of the geomagnetic force, whose partial derivatives

$$\frac{\partial V}{\partial x} = X, \frac{\partial V}{\partial y} = Y, \frac{\partial V}{\partial z} = Z$$

provide the components of the force, is calculated numerically via spherical functions up to the fourth order (namely their 24 coefficients, as no additive constant is used when calculating the potential).

These calculations have been repeated in various ways since 1839, with better observation data, particularly by *Georg von Neumayer* in Hamburg[101]. Here we can see that the approximation with spherical functions up to the fourth order gives a quite good overall picture, which does not become better when going on to curves of the fifth order. *Gauss* showed here, as always, a high sensitivity for practical needs. Also the implementation of the calculation is very interesting in its details.

pp. 308–323, and Vol. XXX (1910), pp. 337–407, of the *Rendiconti del Circolo Matematico di Palermo*].

[101] With the cooperation, in particular, of *Adolf Schmidt*. [More detailed references can be found in the paper of Adolf Schmidt on geomagnetism, *Enzyklopädie der Mathematischen Wissenschaften*, VI 1, 10, n. 20.]

Second Part: Free Geometry of Plane Curves [104]

We now turn to the second main part of this lecture course, which I call "free geometry of plane curves". The adjective "free" indicates that we are not dealing with definitions and developments linked to the choice of a fixed orthogonal coordinate system, as was done so far predominantly for the considerations on the concept of function. To save space, our treatment will be limited to plane curves.

Once again, a dichotomy naturally arises between considerations within precision mathematics and within approximation mathematics. As we have done so far, here too our reflections will be related to the concepts developed for analytic geometry, that is, a geometry that is explicitly based on the concept of number (thanks to the axioms of geometry), unlike synthetic geometry, which – through the same axioms – operates directly on figures. This approach is not caused by any objective necessity; we could also conceive of all the questions that we are going to discuss in terms of synthetic geometry, but, on the one hand, we would loose the outlook onto other domains of mathematics that are not yet considered by synthetic geometers, and, on the other hand, within the literature of this field, studies have been realised mainly in analytic form.

Here the relationship between analysis and geometry turns out to be handled not differently from how it was handled in the first part: We clarify geometric ideas by accompanying them with analytical developments, and we revive analysis by considering geometric figures.

I. Precision Geometry Considerations in Plane Geometry

Theorems About Point Sets

We start with precision geometry considerations. Set theory will prove to be fundamental in the questions we treat. With the following list we recall the most basic definitions needed:

1. The coordinates x, y of a point are numbers according to the modern concept of number, that is, we define them as finite or infinite decimal fractions with a positive or negative sign. **[105]**

2. A set of such points becomes, if it is infinite, object of a detailed mathematical analysis. We may find remarkable differences[102]:

a) An infinite point set (x, y) can be *denumerable* or not. It is called denumerable if it is possible, somehow, to put its elements in a one-to-one correspondence with the natural numbers.

b) An *interval* refers to the totality of points that lie within a rectangle whose sides are parallel to the coordinate axes. If one wants to stress that the points of the border of the rectangle do not belong to the interval, the latter is called an *open interval*. In the other case, in which the points of the border are considered as belonging to the interval, it is called a *closed interval*. For the totality of the points of *any* bordered part of the plane one speaks of a *closed domain* or of an *open domain* (= region) (see p. [85][103]).

c) An infinite point set (x, y) is called *bounded* if the absolute value of the coordinates of all its points remains below a fixed positive number.

d) A point $P(x, y)$ is an *accumulation* point for a given infinite point set if in every interval (domain), however small, of the *xy-plane* containing P there are infinitely many points in the set. The following theorem holds:

Every infinite bounded point set (x, y) has at least one accumulation point.

---

[102] [For the following, the main references are: 1. The already mentioned reports in the Enzyclopädie II C 9 of *Arthur Rosenthal*. – 2. *Jordan, Camille:* Cours d'Analyse, 3rd edition, Vol.1. Paris 1909. – 3. *Hausdorff, Felix.:* Grundzüge der Mengenlehre, 2nd edition, Leipzig 1927. – 4. *Carathéodory, Constantin:* Vorlesungen über reelle Funktionen, 2nd edition, Leipzig and Berlin 1927. – 5. *Hahn, Hans:* Theorie der reellen Funktionen Vol.1. Berlin 1921.]

[103] [for a more detailed analysis of the concept of "domain" see pp. [114]–[116]]

F. Klein, *Elementary Mathematics from a Higher Standpoint*,
DOI 10.1007/978-3-662-49439-4_7

e) If P is an accumulation point for a point set, P itself can belong to the set or not. If a point set contains all its accumulation points it is called *closed*.

f) A point set is called *dense in itself* if each of its points is an accumulation point for the set.

g) A point set is called *perfect* if it is closed and dense in itself.

h) A point set is *everywhere dense* in an interval (domain) I of the xy-plane, if in every subinterval (subdomain) of I there are infinitely many points of the set. For instance, the set of all rational points of the *xy-plane* is everywhere dense in any interval of the plane. If in none of the subintervals (subdomains) of I the set is *everywhere dense*, it is called *nowhere dense* in I.

[106] As you can see, these are all concepts that are per se easy to understand, whose appreciation and understanding, however, emerges only when working on concrete examples.

I want to lead you from the start to the conviction that when studying point sets we are dealing with genuine geometric issues. Therefore, rather than considering – which is what usually happens – examples that seem arbitrarily constructed in the realm of arithmetic, thus recalling the idea of artificiality, we want to use a *geometric principle for the generation of a point set*, which was developed within *the theory of automorphic functions*.

Point Sets Obtained by Inversion with Respect to Two or More Disjoint Circles

We start by considering the *transformation by reciprocal radii* or *circle inversion*.

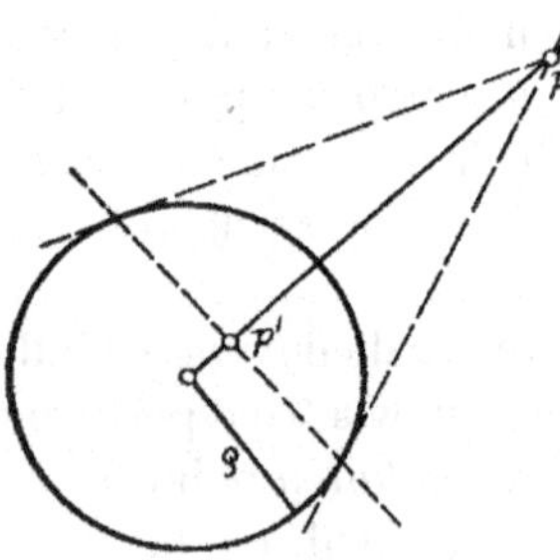

Figure 56

A circle of radius ϱ and a point p external to the circle are given. The law of inversion expressed by the formula $rr' = \varrho^2$ yields then a point p' in the interior of the circle[104], which is called the inverse point of p (see Fig. 56). This transformation

[104] Translator's note: r is the distance between p and the centre, and r' the distance between p' and the centre.

by reciprocal radii is sometimes called "reflection in a circle"; but this has to be understood *cum grano salis*, because this does not have to do with the law of optical reflection by a convex mirror. Only if the circle degenerates into a straight line we have reflection in the optical sense.

Concerning our transformation, I assume as known that it maps circles into circles preserving the measure of the angles that they form when cutting each other – though changing their orientation. We emphasize that the straight line is counted among the circles, simply as a circle with infinite radius. Infinity is seen as a single point; the straight lines are those circles, which pass through the ∞ point. The relation between p and p' can also be defined saying that p and p' are the points common to a family of circles that cut the circle K *orthogonally*. This definition has the advantage of considering only elements that are invariant under any other inversion.

In the theory of automorphic functions an arbitrary number of such transformations are combined together. Take the following example: Consider three disjoint circles K_1, K_2, K_3, and a point p in the common region external to the three circles. We construct the inverse points of p, namely p_1, p_2, p_3 and continue from these points reflecting them again with respect to the given circles, etc. **[107]**

The different points thus constructed are represented in symbols as

$$(p)S_1^{\alpha_1}S_2^{\alpha_2}S_3^{\alpha_3}S_1^{\beta_1}S_2^{\beta_2}S_3^{\beta_3}\ldots,$$

where S_1, S_2, S_3 are the reflections with respect to the circles K_1, K_2, K_3, and $\alpha_\nu, \beta_\nu, \ldots,$ are integer numbers, which denote how many times a transformation has to be performed. As each inversion is an involutory transformation, S_1^2, S_2^2, S_3^2 always represent the identity, and it is sufficient to limit the numbers $\alpha_\nu, \beta_\nu, \ldots$ to the values 0 and 1. The question is: how can we get an image of the point set thus constructed? Which accumulation points does it have?

So, in the following, we are dealing with a point set obtained from a given point p by applying any combination of three given inversions, that is applying to p the whole group of "transformations" that arises from the three "generators". We name the considered points briefly *equivalent points* to p.

If someone objects that the question appears still very artificial, notwithstanding its purely geometric form, we recall that the point set considered has been observed by *William Thomson* and *Bernhard Riemann* already in 1850 while studying *problems of electrostatics*. In fact, when studying the equilibrium of charges on three rotating cylinders with parallel axes (whose normal section becomes our plane figure) the "method of image charges" leads exactly to the generation and to the study of our point set.

At first we want to face the question in the more simple case of only two disjoint circles K_1 and K_2 (see Fig. 57); we call the corresponding inversions S_1 and S_2.

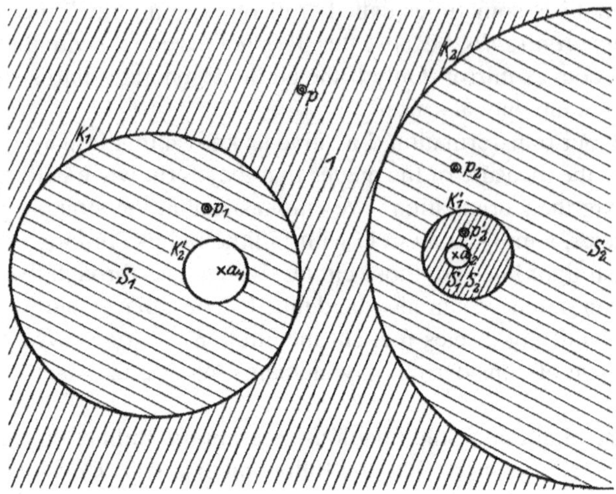

Figure 57

A first artifice is to not let S_1 and S_2 operate on one single point p, but con-
[108] tinuously on all the points belonging to the common region external to the circles, which we denote by 1.

We call the regions obtained through these operations *equivalent* to 1 (and thus also mutually equivalent); we will see that these regions confine gaplessly and without any overlap, so that in the end the whole plane is filled, with the exception of two points a_1 and a_2 that we indicate as *limit points. The overall view on the set of mutually equivalent points is obtained as soon as we gain the overview on these mutually equivalent regions:* Because any point p of the starting region will have in each of the equivalent regions one and only one equivalent point.

For region 1, two new regions result by applying the reflections S_1 and S_2, that we call again S_1 and S_2. S_1 adheres to 1 gaplessly along K_1 and is, for its part, limited in its interior by a new circle, which is called K_2' in the Figure, because it is the image of the circle K_2 (arising through S_1). Analogously the region S_2 lies alongside 1, and has K_1' as its interior limiting circle.

The matter is now the construction of new regions equivalent to 1, and eventually of the totality of these regions. To this purpose we firstly want to reflect S_1 and S_2 in their interior limiting circles, that is S_1 in K_2' and S_2 in K_1'. I state that the new regions, which place themselves smoothly in the "openings" of S_1 and S_2, arise from 1 through simple combinations of the operations S_1, S_2; that is, they belong to the series of equivalent regions we are looking for.

In fact, let us take a look at the reflection in K_1'. Two inverse points with respect to K_1' arise, in virtue of S_2, from two points that are inverse with respect to K_1. Because those are the basis points of a family of circles orthogonal to K_1', and therefore the two points, from which they arise thanks to S_2, are basis points of a family of circles orthogonal to the circle from which K_1' arises thanks to S_2, that is to K_1. The region created by the inversion of S_2 with respect to K_1' can therefore also be generated by transforming, through the operation S_1, region 1 into S_1 and then applying to the latter the operation S_2. The new region must therefore be denoted by S_1S_2 and it effectively belongs to the series of regions equivalent to 1 we are looking for. In the same way the region that arises from S_1 by reflection over K_2' will be denoted by S_2S_1.

The constructions just described exemplify a *principle of continuation*, which we will use to generate again and again new regions equivalent to 1. *Again and again we will reflect each new region (beginning with S_1S_2 and S_2S_1) in its internal limit circle.*

In this manner we obtain, on the one hand, the regions $S_2, S_1S_2, S_2S_1S_2, S_1S_2S_1S_2, \ldots$, (Fig. 57, right) which succeed and include one another gaplessly, **[109]**
and, on the other hand, the analogously succeeding regions $S_1, S_2S_1, S_1S_2S_1 \ldots$, (Fig. 57, left), that indefinitely contract to the two limit points a_1 and a_2.

Now, the beautiful thing is that, in this way, the regions obtained by means of our principle of continuation *exhaust the totality of the mutually equivalent regions.* Indeed, every combination, which can be built from the symbols S_1 and S_2 by repetition, considering that $S_1^2 = 1$ and $S_2^2 = 1$, turns out to be nothing but a mere sequence of alternating S_1, S_2, and that every such sequence turns out to designate one of the regions obtained through our principle of continuation. If the sequence that we might consider finishes with S_2, then the corresponding region is on the right of Fig. 57, otherwise on the left.

Thus, in the figure produced by means of our principle of continuation we can see the *totality* of the mutually equivalent regions, and therefore we also have an overview on the totality of the points equivalent to p. At the same time, we observe that the total figure is inverse to itself not only with respect to K_1 and K_2, but also with respect to any circle representing the boundary between two neighbouring regions, so, for instance, with respect to K_1' or K_2'. The proof can be performed for the circle K_1'. By inversion in K_1' the region S_2 gives rise, as we know, to the region S_1S_2. *The inversion in K_1' can therefore be indicated as $S_2^{-1}S_1S_2$, or $S_2S_1S_2$, which is the same because $S_2^2 = 1$.* If I now want to apply this operation to any region that we produced in this way (which is denoted by any sequence of the symbols S_1, S_2) then I only need to add to this sequence the sequence of letters $S_2S_1S_2$, by which another region existing in our figure can evidently be designated. Here lies the proof. With the theorem just proved we also obtained a *generalisation of the principle of continuation.*

Until now we only used the circle K_1' to reflect the confining region S_2 in it, thus obtaining the analogously confining region S_1S_2. Obviously, we can just as well reflect in K_1' simultaneously *all* the three constructed regions, namely the adjacent regions $S_2, 1, S_1$, thus obtaining at one stroke three new regions, namely $S_1S_2, S_2S_1S_2, S_1S_2S_1S_2$. Then we have in total six adjacent regions, and all of these can be simultaneously reflected in the interior limit circle of $S_1S_2S_1S_2$, obtaining *at one stroke* six new regions, etc. I call this procedure the *extended principle of continuation.*

The relations that we have described above, which can be studied in detail **[110]**
through a figure, also give us the means to manage more difficult cases, where the

inversions have to be combined together unlimitedly with respect to more than two given circles. We start by considering the case of our original task, which referred to three disjoint circles K_1, K_2, K_3.

In particular, we take the circle K_3 as a straight line (see Fig. 58). To obtain a region 1 whose borders are the circles K_1, K_2 and the line K_3, the circles K_1 and K_2 must clearly both lie to the left or to the right of K_3. We apply the inversion with respect to K_1, K_2, K_3 to region 1 and obtain the regions S_1, S_2, S_3, each with two internal limit circles K_2', K_3' etc. Our principle of continuation now requires that we continue to reflect the new regions S_1, S_2, S_3 in the circles that represent their inner limit. We always obtain a new region, again with two internal limit lines. We reflect again in the latter, and so on till infinity. So the whole plane is "simply" covered with infinitely many mutually equivalent regions, or, respectively, with sets of infinitely many mutually equivalent points, with the exception of the limit points, toward which tend the openings of each new constructed region when these become indefinitely small and more and more numerous. The question arises: What else can be reasonably said about the set of the limit points, while the power of imagination already weakens after a few inversions?

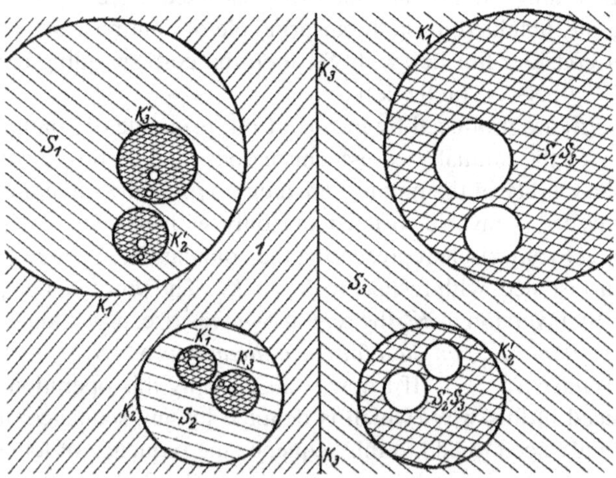

Figure 58

First of all, I again sum up the essential things in three points:

1. We subject not a single point p, but the whole region 1 (the fundamental region) to the transformations.

2. The multiplication of our initial region occurs best when each new region, or
[111] even the whole figure so far obtained, is reflected again and again on the free limiting circular lines, which it has (original and extended principle of continuation).

3. The whole plane is in this way completely covered by infinitely many images of the original region, which confine each other gaplessly, with the exception of the limit points toward which our recursive construction tends.

One could obviously consider also four, five or more initial circles. But there a complication arises, which I now want to describe. Or better, I want to emphasize that the figure with 3 circles still enjoys a certain simplicity.

As is known, three circles have one and only one common orthogonal circle, and this unique common orthogonal circle is mapped into itself by the inversions S_1, S_2, S_3 because it is the only existing orthogonal circle. Therefore it is mapped into itself by all combinations of these inversions. Consequently, all the infinitely many circles that arise by inversion starting from three circles are perpendicular to the fixed orthogonal circle, that is, all the regions that arise from the fundamental region somehow rest on the orthogonal circle: This means that they penetrate it and are symmetric with respect to it. The regions are "aligned" along the orthogonal circle "one next to the other". As a consequence, all the limit points we are speaking of are arranged along the perimeter of the common orthogonal circle. Now, this is highly reassuring in so far as we have, so to say, a fixed reference for the evaluation of the position of the limit points.

If we pass now to four or more circles that do not cut each other and are situated so that they encircle a common region, a common orthogonal circle may exist, though, so that the limit points present the same behaviour. However in general, this is not the case, so that there is no Ariadne's thread to lead us safely through the labyrinth of the equivalent regions and their limit points.

But now we want to assert something about the manifold of the points. First of all, I repeat that a concrete spatial representation of the result is impossible already in the case $n = 3$, even less so for $n > 3$; our imagination already finds its limits at the first inversions[105] and – confronted here with the infinite – Schiller's perennial **[112]**
words also hold for us ("die Größe der Welt"[106]):

Senke nieder
Adlergedank', dein Gefieder!
Kühne Seglerin, Phantasie,
Wirft ein mutloses Anker hie.

But it is all the more admirable that, thanks to the *conceptual* definition of our point set, it is actually possible to assert something about it. It has been a particular merit of *Georg Cantor* (1845–1918) to make this possible; Cantor was the first to show how infinity could be itself the object of mathematical considerations.

---

[105] This is a particularly beautiful example for my claim that the objects of precision geometry lie beyond our imagination.

[106] Translator's note: we found the following anonymous translation at http://www.best-poems.net/friedrich_von_schiller/poem-7828.html

The Greatness of the World:
Eagle-thoughts that aspire,
Let your proud pinions tire!
For 'tis here that sweet phantasy, bold to the last,
Her anchor in hopeless dejection must cast!

Properties of These Point Sets

Suppose that the circle illustrated here (Fig. 59) is a limit circle of our figure, into which all the details of the external figure are reflected. It is clear that the limit points do not fill the whole inner part of the circle, instead, there is space for new regions – at first free from limit points – with new circular borders. We infer that it is not possible to specify any finite region, nor any finite part of a curve, in which the limit points should be everywhere dense. So we arrive to our first conclusion:

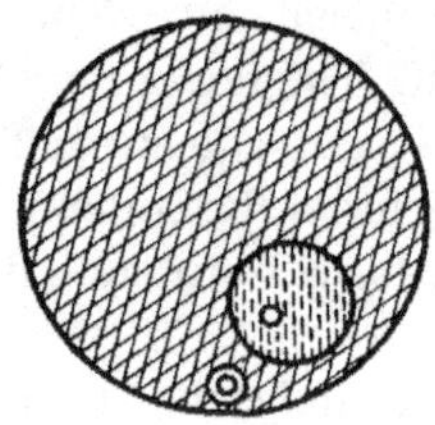

Figure 59

Figure 60

1. *From our continuation principle it results that the set of the limit points is nowhere dense.*

We now select a limit point and place a small boundary circle around it (Fig. 60). It is then clear, that – given that we can reflect the whole external part of the figure in this circle – we can enter into this boundary circle with infinitely many limit points and we can thus come arbitrarily close to our initial limit point, that is, each limit point of our representation is an accumulation point for infinitely many limit points. *The set of limit points is therefore dense in itself.*

Further, we want to investigate whether this set perhaps is also perfect, that is, whether it contains all of its accumulation points. Suppose that p is an accumulation point for limit points. Then we can find limit points in every neighbourhood of p, however small. We then obtain right away a contradiction if we suppose that p itself is not a limit point. In fact, in this case the point p should lie either in the inner part of any of the equivalent regions, which we can call B, or on its interior boundary circle. In the first case it is possible to specify a neighbourhood of p that contains only internal points of B, and therefore contains no limit points, in contradiction to our assumption. In the second case we reflect B in its internal limit circle, thus obtaining a neighbourhood made up of internal and boundary points of the region
[113] B and of its reflected image, which is still free from limit points, in contradiction to our assumption. So we obtain the theorem:

2. *The set of limit points is perfect.*

Theorems 1. and 2., which appear to be incompatible according to our habit formed by the finite, do not exclude each other when we are dealing with, as we are here, *infinite* point sets.

We further search for the *cardinality of the limit points*. Each single limit point is completely characterized by an infinite sequence of the digits 1, 2, 3, which is formed in such a way that two adjacent digits are never equal.

The question is now: How many of such infinite sequences exist?

The cardinality of all these combinations of three numbers is the same as the cardinality of all dyadic fractions, that is, of all fractions that advance with the powers of $\frac{1}{2}$; I will not deepen this issue. So we arrive to the *cardinality of the continuum*.

You see that already with the simplest examples of automorphic figures we obtain together the extremely beautiful properties of the sets.

The theory of automorphic functions leads to figures that are even much more complicated than those we have just considered (see the numerous figures in *Robert Fricke-Felix Klein*: Theorie der automorphen Funktionen Vol. 1, pp. 428 and on, Leipzig 1897). Actually, infinite point sets or other constructions, which geometry investigates, necessarily arise when introducing in geometry the idea of an indefinitely progressing process of construction that follows specific laws. This means that set theory enjoys a fundamental meaning not only for the geometric figures in the theory of automorphic functions, but also for numerous other branches of geometry.

Moreover, all this is of course precision mathematics. The aspect that is characteristic here is completely lost in approximation mathematics.

Third, I am expressing the hope that set theory, as well as it offers its aid to geometry, can also reciprocally receive a new impulse from the latter. Often the examples presented by the experts of set theory appear as somewhat artificial, while we have seen that starting from geometry we can easily reach profound issues of set theory. There, geometry is capable of the most wonderful progresses in its development![107]

[107] [Recently set theory has become crucial in the branch of geometry called topology, both for its exact foundations and for its developments. Moreover, set theory achieved applications in studies on developable surfaces, on minimal surfaces, in the theory of finite continuous groups, in studies on vector fields and on convex structures. The related literature can be found on p. 1013 of the often-mentioned report of the Enzyklopädie by *Rosenthal*. We cannot avoid to complement the considerations of pp. [114]–[127] with a reference to the newest findings in set theoretical topology, namely the discovery of a precise concept of *dimension*, which is sufficiently extended with respect to the previous definitions but which is, nevertheless, in complete accordance with our intuition. This discovery has been made recently, almost simultaneously and independently by *Karl Menger* from Vienna and *Paul (Pavel) Urysohn* (1898–1924) from Moscow; previously, *Henri Poincaré* had shown the way for the solution of this 2000 year old problem (it already appeared at the beginning of Euclid's Elements) in his paper "Pourquoi l'espace a trois dimensions" (Revue de Métaphysique et de Morale 20 (1912), p. 484, reprinted in his "Dernières Pensées"). This new concept of dimension has been the starting point of a new extensive theory, whose results

The Concept of Bi-Dimensional Continuum. The General Concept of Curve

[114] The most obvious next question, which arose again from the field of function theory and to which Weierstraß was the first to need an answer in his lectures, is:

When will a bi-dimensional point set, that is a set on the plane, be called a continuum (or, also, a region)?

The simplest example of what we call a continuum in the plane, is offered by the totality of the points in the interior of a circle or a rectangle, in which we now exclude by convention the boundary points of the set.

[115] We include the two following properties in the *general definition of a plane point continuum*[108]:

1. First of all, we require the "connectedness" of the set: Two points of the set can always be connected through a polygonal path with finitely many segments, whose points all belong to the set.

2. Moreover, it must be possible to construct a circle centred at any point of the set, so that all its interior points belong to the set.

As for the *question concerning the boundary* of such a point set, our knowledge reached so far offers various examples for its delimitation:

1. All the points of the plane, with the exception of one point, can belong to the set.

2. If we recall the example considered on pp. [110]–[113] taken from the theory of automorphic functions, we see that a continuum can also be delimited by an infinite set of points, which need not be dense anywhere.

---

have, however, up to now been only partially published, namely by Urysohn in his voluminous "Mémoire sur les Multiplicités Cantoriennes" (*Fundamenta Mathematicae* Vol. VII. 1925 and Vol. VIII. 1926; continued in the *Verhandelingen* of the Amsterdam Academy 1927), by *Menger* in numerous works in the *Wiener Monatsheften für Mathematik und Physik* and in the Amsterdam Proceedings. Further references for this can be found in the footnotes of the lecture outlined in vol. 36 of the *Jahresbericht der Deutschen Mathematikervereinigung* (p. 9 and on of the pagination in italics). Other short overviews can be found in the "Bericht über die Dimensionstheorie" (*Jahresbericht der Deutschen Mathematikervereinigung* Vol. 35 (1926), pp. 113–150), written by Menger himself, and in the two works: *Paul Alexandroff*, Darstellung der Grundzüge der Urysohnschen Dimensionstheorie, *Mathematische Annalen* Vol. 98 (1927), pp. 31–63, and *Witold Hurewicz*, Grundriß der Mengerschen Dimensionstheorie, ibidem Vol. 98 (1927), pp. 64–88. The definition mentioned is based on the concept of neighbourhood (see *Felix Hausdorff*: Grundzüge der Mengenlehre, 2nd ed., p. 228) and it is the logical abstraction of the intuitively evident fact that, to remove a point with its small neighbourhood from a three-dimensional solid, it is necessary to take out a bi-dimensional surface, from a bi-dimensional surface a one-dimensional curve, and from a one-dimensional curve a null-dimensional set made of two points. This concept of dimension is of course of such a nature that the figures that are n-dimensional in the classical sense (for instance a cube in the n-dimensional Euclidean space or even the n-dimensional Euclidean space itself) are n-dimensional also according to the new definition. The proof of this can be found in the works of *Menger* and of *Urysohn*.]

[108] [In the recent topology, a *continuum* is defined, according to *Cantor*, as a *closed* and connected set consisting of not only one point. The "plane point continuum" defined in the text is conceptually identical to a connected *region* of the Euclidean plane.]

3. Moreover, its boundary can consist of a curve in the usual sense (circle, rectangle, etc.)

4. Other boundaries are, however, possible. The following one was considered by *William Fogg Osgood*[109].

For instance, we can restrict our attention to the positive half-plane (the points above the x-axis) and at certain points we make rectilinear cuts perpendicular to the x-axis (Fig. 61); differently from the other points of the half-plane, we do not consider these as belonging to the region. Now, we think of the number of these cuts as increasing. If these cuts lie everywhere dense in no segment of the x-axis, there still exist stripes of the region that approach the x-axis in between the cuts. We have already understood the existence of such sets, which have the cardinality of the continuum and which are nowhere dense but nevertheless perfect; it is sufficient to remember the limit points on the orthogonal circle of our automorphic figure. So we see that a continuum can also have infinitely many cuts, which are grouped like the points of a set that is nowhere dense, but nevertheless infinite and perfect!

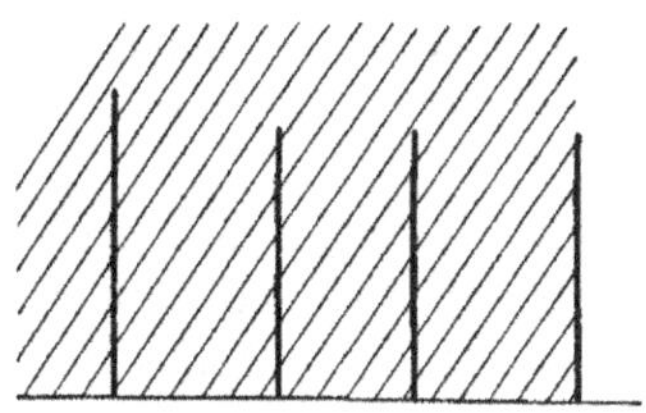

Figure 61

These four examples show us that the boundary of a region (which is usually conceived as a curve) can take many forms, so that – in line with science's state of art today – we must say[110]: **[116]**

A continuum can have boundaries of various kinds, and to briefly define a curve as the boundary of a region is absolutely improper; this happens, of course, only in the realm of precision mathematics, where – because of the modern concept of number – we have to do with idealised structures, in particular with idealised structures of space intuition. *Then how can we define a curve*? To this purpose we use the following approach:

Given a variable t, which can take on all the values of a closed interval $a \leqq t \leqq b$, we set the coordinates of a point as univocal and continuous functions of this auxiliary variable, that is $x = \varphi(t), y = \psi(t)$; one says that the point set thus defined constitutes a curve. The definition of a curve in plain words can be formulated as follows:

A plane curve is a set of points of the plane, which represents a univocal and continuous image of a closed interval on a straight line.

[109] See, for instance, Transactions of the American Mathematical Society Vol. 1 (1900), pp. 310–311.

[110] [See for this point and for the following *Arthur Rosenthal*: Über den Begriff der Kurve. Unterrichtsblätter für Mathematik und Naturwissenschaften Vol. 30 (1924), pp. 75–79. Also see the note 117 on p. 124.]

About the Peano-Curve that Fills a Whole Square

Incidentally, in such a definition of a curve by a parameter t, mechanical interpretations are involuntarily mixed. We can consider t as time (which is also the origin of the usual use of the letter t) and say: While the parameter t runs through the time from a to b, the point $x = \varphi(t), y = \psi(t)$ runs along the curve. Or, stating it differently: The curve is the trajectory of a point that moves continuously during a time interval. All this is easy to understand. It becomes more difficult when we ask for the *characteristics of the* gestalt *relations*, which can arise for a curve defined in this way.

I must draw attention here to a discovery by *Giuseppe Peano*, which he published in 1890[111] and which was geometrically explained by *David Hilbert* in 1891[112]. The discovery is that *a curve defined by univocal and continuous functions* $x = \varphi(t), y = \psi(t)$ *can completely fill an area.* Curves of this kind are called *Peano-curves*; it is not difficult to construct an example of such a curve.

I start by saying what it is *not* meant with such a curve. Someone who has heard of Peano's results could relate them to something known from old times, like the properties of the epicycloids.

[117] We can roll a circle with radius r along a circle with radius R; the ratio $\frac{r}{R}$ is supposed to be irrational. A point arbitrarily fixed on the perimeter of the rolling circle describes an epicycloid that never closes, and consequently it fills in an everywhere dense way the circular region in which it lies, including its boundaries (Fig. 62). This does not mean, anyway, that every point of the region lies on the epicycloid, but only that we can get as close to it as we want if we continue to develop the curve sufficiently far from the starting point. It is not difficult to find on the boundary or in the inner part of the circular region points, which are not reached by the epicycloid.

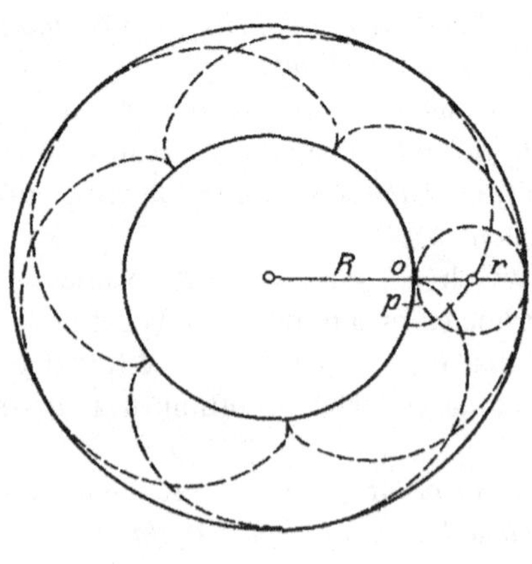

Figure 62

[111] Sur une courbe, qui remplit toute une aire plane, *Mathematische Annalen* Vol. 36 (1890), pp. 157–160.

[112] Über die stetige Abbildung einer Fläche auf ein Flächenstück, *Mathematische Annalen* Vol. 38 (1891), pp. 459–460.

Suppose that at the beginning of the rolling motion the generating point is the point common to the fixed and the moving circle, which has the position denoted by o in our figure. We now consider on the fixed circle the points denoted by p, whose distance from o is an arc of length $\frac{m}{n} \cdot 2\pi R$, where m and n are integer numbers. Since for $\frac{R}{r}$ irrational the length $\frac{m}{n} 2\pi R$ and the circumference $2\pi r$ of the rolling circle are incommensurable, no point p can lie on the epicycloid. Moreover, let us choose a circle k, concentric with the fixed circle, in such a way that successive branches of the epicycloid delimit on k arcs of constant length b, then b must be incommensurable with the circumference u of this circle. If this were not the case, it is easy to understand that the epicycloid would become closed. Suppose that a is a point of the cycloid lying on k, and q another point on k, whose distance from a is the arc $\frac{m}{n} u$, where m and n are again integer numbers: No matter however many times the arc b is carried onto k starting from a, we will never reach the point q. It therefore does not lie on the epicycloid.

Consequently this curve, even though it can be interesting per se, does not realise the proposition of Peano's theorem, because, for Peano, a curve should be constructed that, for a particular value of t, really *reaches* every single point of a region.

Now we come to the geometric explanation of Peano's curve. As an example, we choose a curve that completely fills the closed square-interval $0 \leqq x \leqq 1$, $0 \leqq$ **[118]**
$y \leqq 1$ of the xy-plane.

We define this curve as the limit curve C_∞ of a sequence of curves $C_1, C_2, C_3 \ldots,$ where C_1 is the diagonal of a square, that goes from the origin $0, 0$ to the point $1, 1$ (Fig. 63). We take as parameter t the "arc-length" of C_1 divided by $\sqrt{2}$. As functions $\varphi_1(t)$ and $\psi_1(t)$ we then obtain $x = t$, $y = t$, when t takes on all values in the interval $0 \leqq t \leqq 1$ (see Fig. 64).

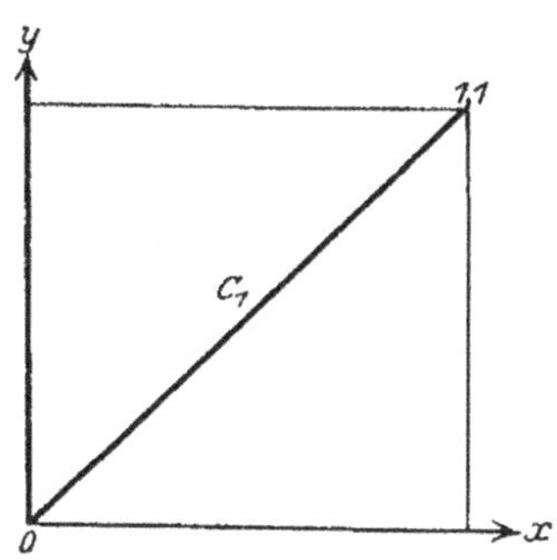

Figure 63

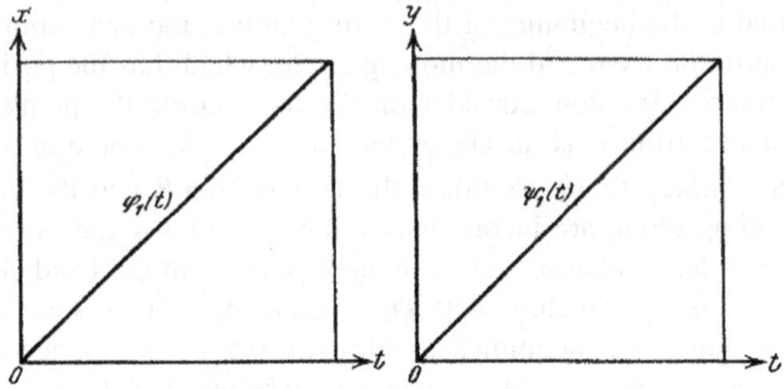

Figure 64

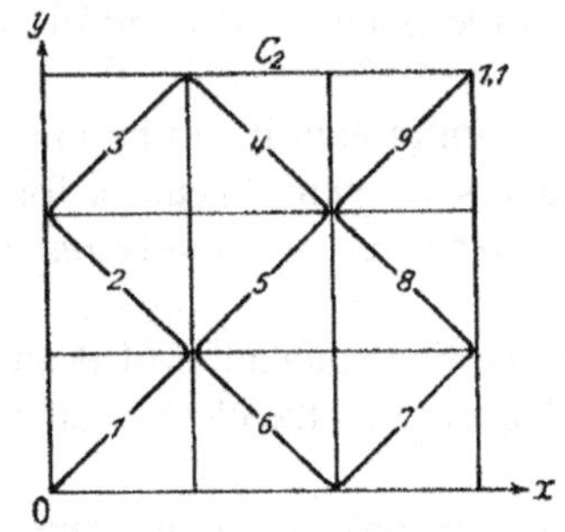

Figure 65

Now we make the decisive step to achieve C_2. We maintain the starting point and the end point of the diagonal unchanged, we divide the square in 9 smaller squares, and take as C_2, instead of the large diagonal, the 9 smaller diagonals of these new squares[113], which are ordered as shown in Fig. 65. The figure reveals roundings at those points where the curve trespasses from one subsquare to another, so as to allow one better to follow the succession of the various parts of the intended curve. One can see that two points in the interior of the large square, namely the corner point common to the squares $1, 2, 5, 6$ and the one common to the squares $4, 5, 8, 9$ are passed twice by C_2. The new curve is obviously three times longer than C_1, has therefore the length $3\sqrt{2}$. If we define the parameter t as the length of the arc C_2 divided by $3\sqrt{2}$, then we obtain for $x = \varphi_2(t), y = \psi_2(t)$ the polygonal path represented in Figs. 66 and 67 (where t runs again from 0 to 1). The curve φ_2 arises then from ψ_2 when we reduce the latter in the ratio $1:3$ and draw it three times
[119] alongside the diagonal φ_1.

[113] Instead of 9 we could take the square of an odd number > 3, without any relevant change.

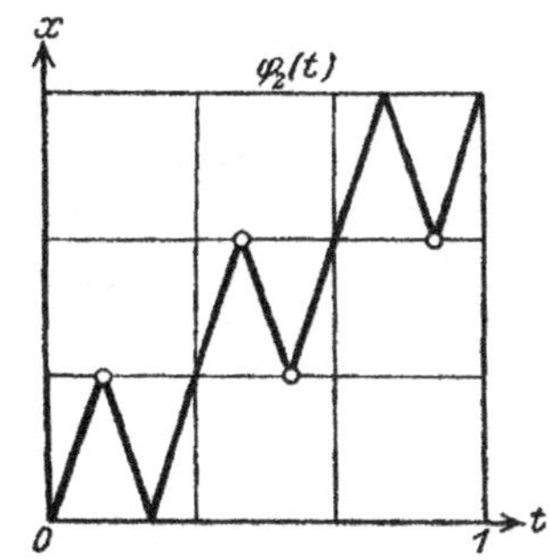

Figure 66

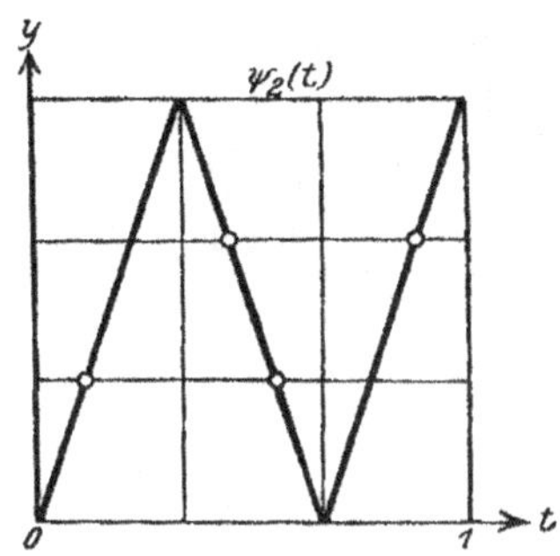

Figure 67

On its part, the curve ψ_2 arises from φ_1 when the latter is compressed to one third of the original square, laterally, and the obtained figure, resp. its image obtained by reflecting it on a vertical straight line, is juxtaposed three times in alternating order.

We now repeat the transition from C_1 to C_2 in each of our nine minor squares and proceed indefinitely in this way.

In the same manner in which we passed from C_1 to C_2 by dividing the square into nine parts, we now pass from C_2 to C_3 by a further subdivision of the individual squares into nine parts, and substitute every former diagonal with a polygonal path, which either represents the original curve C_2, or a mirror image in a reduced scale (Fig. 68).

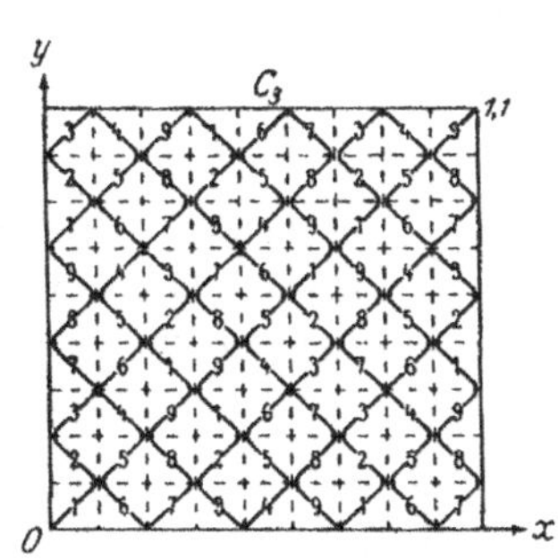

Figure 68

We proceed in a similar way to obtain C_4, and so on.

If, as before, we now want to draw $y = \psi_3(t)$, we divide the interval $0 \leqq t \leqq 1$ and also the interval $0 \leqq y \leqq 1$ into 9 parts, thus obtaining a curve $\psi_3(t)$, which is composed of 27 segments, each of which is again composed of 3 single small pieces (components), so that we have in total 81 components, corresponding to the 81 diagonal pieces of the curve C_3 (Fig. 69[114]).

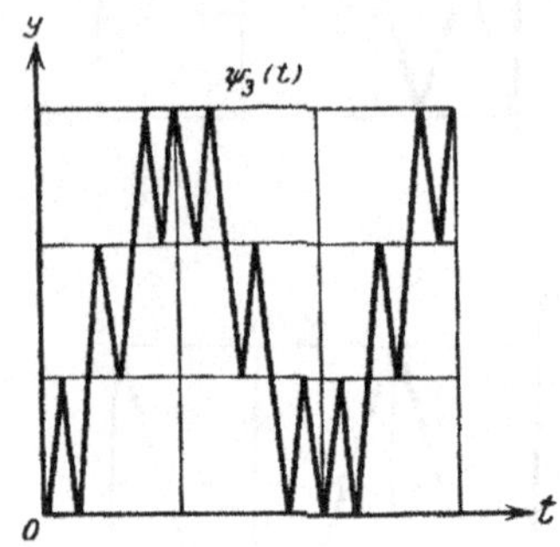

Figure 69

[120] If, by the way, we compare ψ_3 with φ_2 we can say that ψ_3 grows out from φ_2 like ψ_2 from φ_1.

Again, the curve φ_3 arises by applying three times the curve ψ_3 just drawn, reduced in a scale of 1: 3, alongside the diagonal φ_1.

But we also can deduce φ_3 from φ_2 and ψ_3 from ψ_2. In general, the following holds:

To pass from φ_n to φ_{n+1} (or from ψ_n to ψ_{n+1}, for $n \neq 1$) we must split at first every rectilinear side of φ_n (or of ψ_n) into three parts of equal length, then replacing every rectilinear part by a three-pronged hook leaving the common initial and final points (Fig. 70).

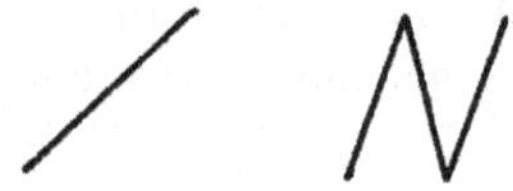

Figure 70

This law of progression for the curves φ_n and ψ_n reminds us of the *partial curves of the Weierstrassian function*, with the difference, however, that – instead of the oscillations of the sines – here rectilinear jagged curves are superposed. The height of

114 [In Fig. 69 you can think of each of the nine drawn subsquares as divided like in Fig. 68.]

the jags becomes more and more insignificant as n increases, but the width decreases much more rapidly, so that the *shape* of the jags becomes steeper and steeper[115].

Now, I try to describe the limit curves $\varphi_\infty, \psi_\infty, C_\infty$ more in detail.

Let us call corners[116] the initial and the final points of φ_n and ψ_n, and also those points where the various rectilinear parts of these curves touch each other (φ_3 and ψ_3 have thus each 82 corners), so we have in these corners consistently points that also belong to φ_{n+1} and ψ_{n+1}, and to all other successive curves. The curves φ_∞, respectively ψ_∞, are filled everywhere densely with such corners. Thus we obtain a complete image of the continuous curves φ_∞ and ψ_∞, assuming to have constructed, for increasing n, the respective corners of the curves φ_n and ψ_n, and considering that these belong in their totality to the limit curve.

Going back now to the curves C_n, one finds that the break-points, in which the diagonals join each other at a right angle, correspond to the corners of the curves φ_n and ψ_n, with the exception of the initial and the end points; the break-points are (as far as they are break-points internal to the initial square) multiple points of the C_n, and will become such for C_∞, as well. The corners of *one specific* C_{n_0} remain unchanged in all the successive C_n $(n > n_0)$, with the only difference that more and more corners fall in between.

The fact that the curve C_∞ *completely fills the square* $0 \leqq x \leqq 1, 0 \leqq y \leqq 1$ *can be proved in the following way*: To construct C_2 we divided the square into nine **[121]**
equal subsquares, to construct C_3 we divided each of these squares again into nine equal squares, and so on.

Each of these subsquares corresponds to a subinterval of the segment $0 \leqq t \leqq 1$. Now we can consider an arbitrary point Q on the boundary or in the interior of the initial square as limit point of an infinite sequence of nested squares, whose diagonals J_n decrease indefinitely toward zero with increasing n. This nesting of squared intervals corresponds, as one learns by taking a look at our figures, to a nesting of line-intervals on the t-axis, which defines a point P in the interval $0 \leqq t \leqq 1$. Thus, there exists at least one parameter t_0 that yields the point Q. This must, therefore, belong to the Peano-curve C_∞. Finally we can easily prove, based on our figures, that for each value t there is only *one* point of the Peano-curve, and that therefore the mapping of the segment $0 \leqq t \leqq 1$ onto the square $0 \leqq x \leqq 1$, $0 \leqq y \leqq 1$ produced by the curve is univocal. For the proof we introduce the following counting of the subintervals:

I. Counting of the subintervals of the interval $\overline{01}$.

We count the intervals that arise from the first subdivision T_1 from the left to the right and call them

$$\delta_1, \delta_2, \ldots, \delta_9.$$

---

[115] [The continuity of the functions φ_∞ and ψ_∞, is proved on p. [122] – despite their continuity, they are not differentiable, and thus "*Weierstraß functions*". See *Ludwig Bierbach*: Differentialrechnung, 2.nd ed. Leipzig 1922, pp. 107–111].

[116] Translator's note: literally "hinge-points" (Gelenkpunkte).

With the second subdivision T_2 each δ_v is divided into nine equal parts, and counting the new subintervals again from left to right we obtain in δ_1 the subintervals

$$\begin{array}{ll} & \delta_{11}, \delta_{12}, \ldots, \delta_{19}, \\ \text{in } \delta_2 & \delta_{21}, \delta_{22}, \ldots, \delta_{29}, \\ & \ldots\ldots\ldots\ldots\ldots \\ \text{in } \delta_9 & \delta_{91}, \delta_{92}, \ldots, \delta_{99}. \end{array}$$

We conceive of the process of dividing and counting being continued indefinitely.

If a point P of the segment $\overline{01}$ is not an interval endpoint in any subdivision, then it is determined by one, and by only one, sequence of nested intervals. But if a point P is an endpoint of a subinterval, then it is defined by two sequences of nested intervals (with the exception of the points 0 and 1). Among these, one contains – from a certain subdivision on – intervals that touch P only coming from the right, and which therefore end with 1; the other one contains intervals that touch P coming from the left, and therefore end with 9. For instance, the right end-point of δ_{12} is defined both by the nesting

$$\delta_1, \delta_{12}, \delta_{129}, \delta_{1299}, \ldots$$

and by the nesting

$$\delta_1, \delta_{13}, \delta_{131}, \delta_{1311}, \ldots$$

[122] II. Counting of the subintervals of the square on $\overline{01}$.

We number the nine subsquares obtained by the first subdivision from 1 to 9, as shown in Fig. 65, and call them

$$\eta_1, \eta_2, \ldots, \eta_9.$$

Then we denote with

$$\eta_{v1}, \eta_{v2}, \ldots, \eta_{v9} \quad (v = 1, 2, 3, \ldots, 9)$$

the squares obtained by the second subdivision.

In the enumeration of these 81 squares we pay attention to the following rules: a) without passing twice on a subsquare and without jumps, we first assign the numbers 1 to 9 to the squares of η_1 and then to those of η_2 as well, and so on. b) Also when passing from η_1 to η_2, and in general from η_v to η_{v+1}, we do not jump, so that the subsquare of η_v with number 9 and the subsquare of η_{v+1} with number 1 have a common side (see Fig. 68).

The processes of division and counting are considered as continuing indefinitely.

It is possible to see that: To each linear interval $\delta_{\alpha_1,\alpha_2,\ldots,\alpha_v}$ corresponds only one square interval $\eta_{\alpha_1,\alpha_2,\ldots,\alpha_v}$, and consequently to each sequence of nested intervals (δ) only *one* sequence of nested intervals (η). So, if a point on $\overline{01}$ is not an endpoint of a subinterval, it corresponds to only *one* image point Q inside the square. This is surely the case also when P lies in 0 or in 1. If, instead, P is an endpoint of a subinterval and if (δ) and (δ') are its two nested sequences, then – according to the numbering rule b) – the corresponding nested sequences of squares (η) and (η') also fix only *one* point. Thus it is proved that to each value of the parameter t corresponds only *one* point of the Peano curve.

We finally provide the proof of the continuity of the functions $x = \varphi_\infty(x)$, $y = \psi_\infty(x)$ and therefore of the continuity of the Peano curve. Suppose that P and P' are two points on the segment $\overline{01}$, t and t' their parameter values, (δ) and (δ') their two sequences of nested intervals. If t, respectively t', is an endpoint of a subinterval, then (δ), resp. (δ'), always denotes the *left* sequence; for the moment we exclude that t or t' lie at the origin. The nested intervals (δ) and (δ') may have n common subintervals. Now, if Q and Q' are the image points of t and t' on the Peano curve, (η) and (η') are their squared nested intervals, then the number of the squares common to (η) and (η') is n, as well. If we let t' approach t indefinitely, then n increases indefinitely and at the same time the minimal length of the intervals common to (δ) and (δ') tends to zero indefinitely. But, as n increases the diagonal of the subsquares decreases indefinitely, so that it results:

$$\overline{QQ'} \to 0 \quad \text{for} \quad P' \to P.$$

The continuity of the three curves $\varphi_\infty(x), \psi_\infty(x), C_\infty$ is thus proved for all the points of the interval $\overline{01}$ (the proof for the case in which P lies in the origin is trivial). **[123]**

A More Specific Concept of Curve: The Jordan Curve

Now we pass to the following manifest question:

The Peano curve does not recall what is said to be a curve in every-day life. *How do I need to restrict the definition of a curve $x = \varphi(t), y = \psi(t)$, or, which properties do I need to postulate for a thus defined curve in order to obtain a behaviour that is analogous to that of an empirical curve*? The answer to this question was given by *Camille Jordan* in his *Cours d'analyse* in an easy manner:

Our Peano-curve has infinitely many break-points and therefore also infinitely many double points (a deeper reflection shows that there are even infinitely many triple and quadruple points); the values of t in the interval $0 \leqq t \leqq 1$, which lead to a multiple point, lie everywhere dense. *Camille Jordan* requires then that the curve defined by $x = \varphi(t), y = \psi(t)$ does not have multiple points inside the interval of definition, that is, that there do not exist two or more values $t_1, t_2, \ldots, (a < t_1 < b, a < t_2 < b, \ldots)$ for which we have simultaneously

$$\varphi(t_1) = \varphi(t_2), \quad \psi(t_1) = \psi(t_2), \ldots.$$

This condition is not required for the endpoints a and b of the interval. If it is also satisfied by these points, then the image of the curve $x = \varphi(t), y = \psi(t)$ is called an *open Jordan curve*. But if $\varphi(a) = \varphi(b)$ and $\psi(a) = \psi(b)$, the initial and the end-point coincide, and the curve is called a *closed Jordan curve*. An open Jordan curve is therefore an invertible, univocal and continuous image of a segment, a closed Jordan curve is an invertible, univocal and continuous image of a circle.

Now a fundamental theorem in analysis holds:

Every closed Jordan curve divides the plane into two connected regions, of which it constitutes the common boundary.

To clarify that this theorem is not obvious, I observe: *The paradoxical aspect of the Peano curve is not the concept itself, but rather the terminology used, namely the fact that we call it a "curve" in a more general sense than allowed if we want to maintain the analogy with an empirical curve.* To clearly underline the contrast implied in Jordan's developments, I want to speak initially not of a Jordan curve but of a *Jordan point set*, for which we will then show that it conforms to the usual curves of empirical geometry with respect to the relations of connectedness and divisibility. In fact, if we briefly say: Every closed curve defined by $x = \varphi(t)$, $y =$
[124] $\psi(t)$, which satisfies the Jordan conditions, divides the plane into an external and an internal region, this sounds obvious.

The reason for this is given by the fact that the term "curve" is here used involuntarily with a double meaning. One should better say more precisely: Starting from an empirical domain, where it is evident that every closed curve divides the plane in an inner and in an outer region, we should ask:

How should we restrict, in an ideal domain, the definition of a point set $x = \varphi(t)$, $y = \psi(t)$ so as to obtain an analogous theorem[117]? The answer is: *To this purpose $x = \varphi(t)$, $y = \psi(t)$ have to satisfy Jordan's conditions.* Afterwards, one might call *curve* such a point set! In this observation lies the meaning of the

[117] [The new concept of dimension mentioned on p. [114] allows to solve the problem of reaching a reasonable definition of curve: *A curve is a compact, one-dimensional continuum.* And a continuum K is called compact if every infinite subset of K has an accumulation point, which – since a continuum is a closed point set – obviously belongs to K. The *non-compact* one-dimensional continua (parabolas), which the naïve intuition considers anyway as curves, can be subordinated to the "general" curves just described, by adding of the point ∞. This concept of curve, under which – in agreement with intuition – point sets like Peano's are not considered to be curves, is the generalization of the concept of Cantor's curve, understood as a bounded continuum of the Euclidean plane, which does not have internal points. References on the theory of curves are: *Karl Menger*, Grundzüge einer Theorie der Kurven. *Mathematische Annalen* Vol. 95 (1925), pp. 277–306; and also other papers published in the same journal, in the *Amsterdam Proceedings*, in *Fundamenta Mathematicae*. The related studies of Urysohn are presented in part 2 of the "Mémoire sur les Multiplicités Cantoriennes" (*Verhandelingen of the Amsterdam Academy* 1927). See also *Paul Alexandroff*: Über kombinatorische Eigenschaften allgemeiner Kurven. *Mathematische Annalen* Vol. 96 (1926), pp. 512–554. Every curve in a Euclidean space of dimension at least two is the boundary of a region, namely the one complementary to the curve; however, simple examples show that the reverse does not hold. We observe, moreover, that in the Euclidean plane – in a sense that is extended with respect to the old one – *closed* curves are defined as boundaries of regions that form bounded continua, they divide the plane in at least two regions, and are the common boundary of all these subregions. The old closed curves (of *Arthur Schoenflies*) are characterized, among these general closed curves, as being "regular" because they divide the plane in exactly two regions (see the previously mentioned work of *Paul Alexandroff*)].

theorem. As for the *proof*, I mention briefly only the main aspects, because the details that need a deeper reflection can be better read in Jordan's *Cours d'analyse*.

Let us think of a point set $x = \varphi(t), y = \psi(t)$ that corresponds to Jordan's conditions. Camille Jordan begins with the construction of an infinite sequence of rectilinear polygons $P_1, P_2, P_3, \ldots$, where each polygon encloses the preceding one, and where no one contains in its interior a point of the set; then he constructs another unlimited sequence of polygons $P_1', P_2', P_3', \ldots$ so that in their exterior no **[125]**
point of the set can be found, and where each polygon excludes the preceding one (Fig. 71).

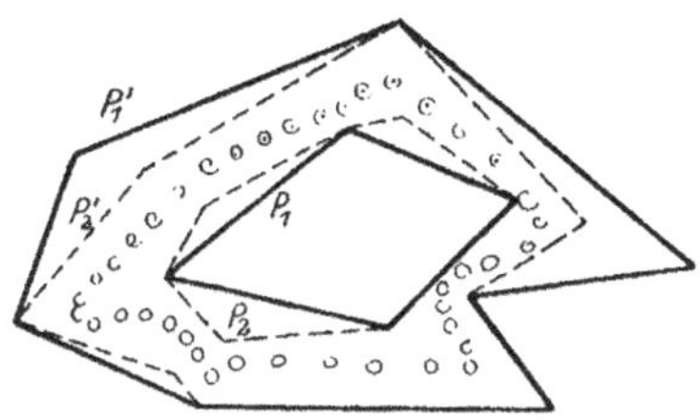

Figure 71

The fact that a rectilinear polygon divides the plane into two regions is considered by Jordan as known from elementary geometry, but it was actually proved later (*Jordan's theorem for polygons*)[118].

After this general consideration, Jordan shows further that every point not belonging to the given set will enter into one of the polygons P resp. P', and finally that the points of our set are left over as accumulation points of the two sequences of polygons P_ν and P_ν'.

The guidelines for this proof are given by the intuition of empirical curves. But every single argument has to be logically based on axioms, and the core reveals to be the modern concept of number. The result can also be expressed as following: *What holds for the perimeters of the polygons, also holds for the limit perimeters.*

Jordan's theorem can be put in parallel to the theorem mentioned at the beginning of the lecture, which states that a continuous function between two of its values also reaches all the intermediate values. In the same way the closed Jordan curve erects a gapless boundary between its interior and exterior.

It is clear from what we have said that we can assign the name *curve* to our Jordan's point set, as regards its relations of connectedness and divisibility. But does it present analogies with the usual curves also to other extents[119]? In other words:

[118] [See *Béla von Kerékjarto*: Vorlesungen über Topologie 1. Berlin 1923; and on new proofs of the Jordan curve theorem see the indications by *Georg Feigl*, Über einige Eigenschaften der einfachen stetigen Kurven, part I, *Mathematische Zeitschrift* 27 (1927), p. 161.]

[119] [A certain number of concepts, like "accessibility", "Unbewalltheit", "local connectedness", "irreducible continuum", which play an important role in modern topology for the treatment of the concept of curve, are not treated here; for what concerns their study we refer to the article of Rosenthal in the Enzyklopädie and to the afore mentioned book of *Kerékjarto*.] Translator's note: the concept of "Unbewalltheit", never became widespread; it corresponds to a property of Jordan

Can we speak, in the case of the Jordan curve, of arc length, tangent, and radius of curvature? Or, which restrictions must be added for φ, ψ for this to become possible? (Of course, the point set of precision geometry needs to be based on exact definitions rather than on the approximated perceptions from the empirical field).

We first face *the question of the arc length of a Jordan curve.*

[126] We divide the finite interval $a \leqq t \leqq b$ in some manner in subintervals $\Delta_1 t, \Delta_2 t, \ldots$ In correspondence to these increments we obtain for x and y the increments $\Delta_1 x, \Delta_1 y;\ \Delta_2 x, \Delta_2 y, \ldots$ We connect the corresponding points of the Jordan set with straight lines, so that the length of the consecutive segments becomes

$$\sqrt{\Delta_1 x^2 + \Delta_1 y^2}; \quad \sqrt{\Delta_2 x^2 + \Delta_2 y^2}; \ldots$$

We assign an arc length to our point set if the perimeter of the constructed polygons, that is the sum

$$\sum_{\nu} \sqrt{\Delta_\nu x^2 + \Delta_\nu y^2}$$

tends to a finite limit, independently from the kind of subdivision of the interval, when we divide $a \leqq t \leqq b$ in more and more indefinitely small parts.

That this is not necessarily so can be seen from the example of the Peano curve, where our question concerning the arc length – after having represented the curve by means of a parameter t – is not touched due to the occurrence of the double points. Here, for C_1, we have the length $\sqrt{2}$, for $C_2 3\sqrt{2}$, for $C_3 9\sqrt{2} \ldots$, for $C_n\ 3^{n-1}\sqrt{2}, \ldots$, thus it is clear that the length of the rectilinear polygons C_n, each of which fits into the corners of our finite point set, tends by no means to a finite limit as the number of sides increases, but it grows beyond any limit.

Now, again, I will give, without providing a proof, the conditions for φ and ψ necessary to assign an arc length to the Jordan point set, only highlighting the main point.

In order to do this I need to introduce a concept much used in modern function theory, namely that of a *function of bounded variation*. A function $f(x)$ is of bounded variation in the interval $a \leqq x \leqq b$ if for every subdivision of the interval $a = x_0 < x_1 < x_2 \cdots < x_{n-1} < x_n = b$ the sum

$$|f(x_1) - f(x_0)| + |f(x_2) - f(x_1)| + \cdots$$

remains below a fixed finite bound A.

We ask again: Do we know an example of a function that is not of bounded variation? I recall the two functions φ and ψ and their curves we constructed as limits of jagged polygonal chains, for the example of the Peano curve. Here we can

regions. For two dimensions it may be formulated as follows: Let J be a closed Jordan curve, I the internal and E the external region. Two points Q and R of J can always be joined by an open Jordan curve belonging entirely to I and by an open Jordan curve belonging entirely to E. Let P be a third point of J and c an arbitrary circle round P. Now the "Unbewalltheit" says that if Q and R are chosen close enough to P, the joining curves may be kept entirely inside c.

immediately see geometrically that the sum $\sum_{1}^{n} |f(x_\nu) - f(x_{\nu-1})|$ grows beyond **[127]** any limit as n increases.[120]

For what concerns the arc length of a Jordan curve, Camille Jordan proves that it results – in the sense of our definition – if and only if φ and ψ are of bounded variation.

Other Limitations to the Concept of Curve: The Regular Curve

If we further ask when a Jordan curve has a *tangent*, or a *circle of curvature*, we can surely give a sufficient condition: Indeed, to obtain that our point set has a tangent or a radius of curvature in each point of a given interval, it is sufficient that φ and ψ are differentiable twice in the interval[121].

When all our exposed restrictions:

1. φ and ψ are continuous in the interval
2. without double points
3. φ and ψ are of bounded variation
4. are differentiable $\nu (\nu \geqq 2)$ times

are satisfied for our point set, then we will designate the Jordan curve without more ado as a *curve*, in consideration of the analogous properties of the empirical curve. However, if we want to emphasize the difference with respect to the Peano curve or to similar entities, then we will call it a *regular* curve. Or, more precisely: we call the so defined point set first of all a *regular piece of a curve. A regular curve is then composed of a finite number of such pieces* and can therefore again have (a finite number of) double points.

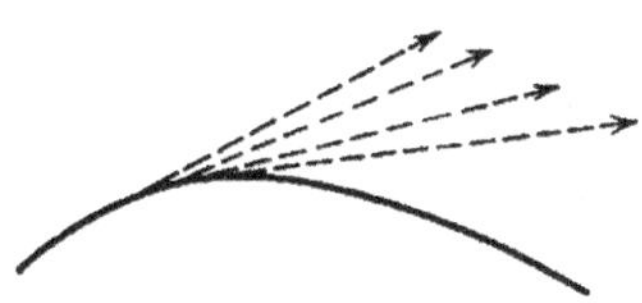

Figure 72

Moreover, I add that one often calls a regular piece of curve what we formerly called "smooth", hence a piece of curve that has a tangent in every point and such that the tangent rotates with continuity upon progressing continuously along the curve (Fig. 72), which is a weaker condition than the existence of a radius of curva-

[120] [Functions of non-bounded variation now have become important for theoretical physics (Theory of white light) thanks to the research of *Norbert Wiener*. See *Norbert Wiener:* Verallgemeinerte trigonometrische Entwicklungen. *Nachrichten der Gesellschaft der Wissenschaften zu Göttingen* aus dem Jahre 1925 (Berlin 1926), pp. 151–158]

[121] This is *not necessary*, because we can substitute a parameter t, for which this is the case, with another parameter $t_1 = f(t)$, for which this is not the case. We have to simply introduce $f(t)$ as a function that is monotonous and continuous but not differentiable.

ture. We will anyhow allow a certain flexibility in the definition of a regular curve,
[128] in that we will adapt to the circumstances our conditions on the first and the higher derivatives of φ, ψ.

I also recall the corresponding developments in the first part of this lecture, which differ from the present ones only because at the beginning we restricted ourselves to point sets which could be represented in the form $y = f(x)$. While we now speak of a *regular curve*, according to the expression we chose now, there we spoke of a "reasonable" function.

Approximation of intuitive Curves by Means of Regular Functions

We ask further: What is the relation between the newly defined entities of precision mathematics and what we simply call "curve" in the empirical field? To avoid ambiguities of the language, due to the fact that we use the term curve in both fields, and, on the other hand, for brevity, I will from now on use the term *idealised curve* within precision mathematics, possibly with the attribute "regular", and speak of *empirical curve* within the empirical field.

I maintain:

We can always associate to an empirical curve a regular idealised curve, that coincides with it according to all the relevant properties exposed before for the curves $y = f(x)$, namely as far as this is at all possible given the restricted exactness of the empirical field.

The proof can be based on the assumption that it is possible to divide an empirical curve into a finite number of pieces, which – in a suitable coordinate system – can be represented by the formula $y = f(x)$ with sufficient precision, also with respect to their slope and possibly their curvature (see the considerations on pp. [51]–[52]).

So we have a finite number of coordinate systems

$$x_1y_1, \quad x_2y_2, \ldots, \quad x_ny_n,$$

and in each of them a regular piece of an idealised curve

$$y_1 = f_1(x_1), \quad y_2 = f_2(x_2), \ldots, \quad y_n = f_n(x_n)$$

these pieces join each other (thanks to the reciprocal position of the coordinate systems) at the initial point and at the end point, so that we have a connected whole. Now we introduce a uniform coordinate system XY and a parameter t, whose interval of reference we divide into n subintervals. If t lies in the first interval, we call it t_1, if it lies in the second t_2, and so on.

The trick consists in merging the n formulas

$$y_1 = f_1(x_1), \quad y_2 = f_2(x_2), \ldots, \quad y_n = f_n(x_n)$$

into only one pair of formulas

$$x = \varphi(t), \quad y = \psi(t)$$

so that for $t = t_1$ the piece $y_1 = f_1(x_1)$ is represented, for $t = t_2$ the piece $y_2 = f_2(x_2), \ldots$ **[129]**

At the same time φ, ψ should be differentiable in t, in these subintervals, as many times as f_1 in x_1, f_2 in $x_2, \ldots$. But this can be obtained in a very simple manner if we put

$$t_1 = \alpha_1 x_1 + \beta_1, \quad t_2 = \alpha_2 x_2 + \beta_2, \ldots$$

and determine the constants α, β so that the end-value of t_1 (which corresponds to the left or to the right end-value of x_1) coincides with the initial value of t_2 (which corresponds to the left or to the right end-value of x_2), equally the end-value of t_2 coincides with the initial value of t_3, etc.

The development sketched out here lets open the possibility that the different pieces of the idealised curve $y_1 = f_1(x_1), y_2 = f_2(x_2), \ldots$, which we strung together, have at the common points *different* slopes, or curvatures. Consequently, the given empirical curve may have a finite number of points, at which the "empirical" slope or curvature changes with a jump. The resulting idealised curve $x = \varphi(t), y = \psi(t)$ will then be affected, of course, by the corresponding singularities.

We might add to the previous considerations the inverse question: *Which properties of the regular idealised curves* $x = \varphi(t), y = \psi(t)$ *can be taken into consideration in the empirical field*?

Evidently all those properties that remain unchanged when we arbitrarily change φ, ψ within given limits, that is, those properties belonging to *approximation mathematics*. We can transform this statement into an intuitive form by imagining, around any point of the curve $x = \varphi(t), y = \psi(t)$, a circle with a small radius ϱ. The totality of the circular areas thus constructed will cover a certain *stripe* of the plane, and it are the figurative properties of the stripes thus defined, and not those of the ideal curve itself, which yield meaning in the empirical field.

Perception of Idealised Curves

Now it is the moment – thanks to the knowledge gathered in the meantime – to come back to the question of the *precision of our spatial perception*, which we raised at the beginning of this lecture course.

Suppose we define, with respect to a coordinate system xy, on the one hand, a circle and, on the other hand, a Peano curve by the corresponding formulas $x = \varphi(t), y = \psi(t)$. Does there exist in principle a difference in our ability to perceive the two newly defined entities spatially? I do not believe so. In both cases we can mentally determine the position of single points of the considered entities by fixing our attention on single values of t, but always only with restricted exactness; **[130]**

we can also imagine, with restricted exactness, the slope of the line connecting two adjacent points, that we suppose fixed; but in none of the two cases can we reach, with our power of imagination, the ideal entity. *The mathematical considerations, by means of which we dominate the ideal structure, are animated and guided by space intuition, but in the end they are based on the principles expressed by the formulas* $x = \varphi(t)$, $y = \psi(t)$, *which are in turn based on axioms (which are also beyond our imagination).*

This is my theory of 1873[122]. I would like to hear the opinion of physiologists and psychologists on this topic. A precondition should only be that they had reflected upon and appropriated the modern developments of precision mathematics, somehow up to the extent of this lecture course. Unless one had deeply reflected upon various examples of idealised curves without derivatives, like the Weierstraß curve or the Peano curve, it is impossible to engage in philosophical discussions about the difference between these idealised curves and the idealised curves having a derivative, which are typically only studied.

Classification of Idealised Curves: Analytical and Algebraic Curves. Geometrical Construction of the Latter as Proposed by Graßmann

Now we turn to a brief discussion of the *most important types of idealised curves*, which privileges a precision mathematics treatment.

1. *When do we call a regular idealised curve analytic*? That is, which restrictions must be posed to the definition of the regular idealised curve, in order to call it analytic? In the following I will call a regular analytic idealised curve briefly an analytic curve.

We will say: A regular idealised curve is called analytic if x *and* y *can be represented by convergent power series in* t[123].

And why do mathematicians deal almost exclusively with analytic curves, even though they have known for a long time that these do not at all exhaust all curves[124]?

The reason lies in the beautiful properties of analytic curves: first of all, they can be generalized to a complex $t = u + iv$. Thus the whole function theory (of complex variables) comes into play (the power series converge in a domain of
[131] the uv-plane). In particular, the "analytic continuation" allows for new domains of definition, and therefore to progress, according to the principles of mathematics, to new pieces of curves, as far as these domains contain a part of the real axis. I conclude by saying: An analytic curve is the collection of all points that are yielded by the power series $x(t)$, $y(t)$ and their analytic continuation, it represents a whole

[122] See the quotation on p. [2].

[123] [As is well known, this definition is equivalent to the one given on p. [62] for the analytic function].

[124] [This observation is today not as appropriate as it was at the time of the lecture].

ruled by mathematical principles, and its complete behaviour is entirely determined by those power series $x(t)$, $y(t)$, which need to converge only within an arbitrarily small region. *The regularity thus defined – the fact that the smallest piece determines the whole – is what makes the analytic curve so popular in mathematics.* For the sake of precision I am adding:

An analytic curve can also have singular points of the most varied kinds; one can obtain them supposing that in the series expansion for φ and

$$x = \varphi(t) = a_0 + a_1 t + a_2 t^2 + \dots,$$
$$y = \psi(t) = b_0 + b_1 t + b_2 t^2 + \dots,$$

the terms of degree 0 until $n - 1$, included ($n \geqq 2$), vanish, and also considering what happens for $t = 0$.

2. We now proceed in making restrictions to the concept of curve, by passing from the analytic curves to the algebraic ones.

The expression already used implies that *an analytic curve $x = \varphi(t), y = \psi(t)$ is called algebraic if φ and ψ satisfy identically an algebraic equation $F(\varphi, \psi) = 0$.*

However, the inverse proposition also holds: Given an algebraic equation $F(\varphi, \psi) = 0$, it is possible to introduce for the neighbourhood of any point x_0, y_0 an auxiliary variable t in such a way, that we obtain $x = \varphi(t), y = \psi(t)$, where φ and ψ are convergent series expansions developing according to powers of t (uniform approximation). I do not prove this, but I mention this theorem, without going into deeper detail, because I am only interested in giving a general overview of this topic.

The further question is now: Is it possible to give, besides this formal definition of the algebraic curve, a more factual one, that is, a definition that can be understood intuitively? This is in fact possible from the point of view both of function theory and of geometry. First of all from the point of view of *function theory*: We construct the function $y = f(x)$ by eliminating t. The question is whether the algebraic function $y = f(x)$ has particularly characteristic properties regarding the series expansion for the individual values of x, through which we distinguish the algebraic function from other analytic functions. Of course, here we need to consider x as a complex variable, because only in this case the statements of function theory find their natural expression.

Indeed, y can now be defined – from the point of view of function theory – as **[132]**
an algebraic function of x by requiring: a) that for the individual value x there are only a finite number P of values of y, and b) that y can be expanded in powers of $(x - x_0)^{\frac{n}{P}}$ in the neighbourhood of any place x_0 without yielding an infinite number of powers with negative exponent (this excludes so called essential singularities). A more precise explanation is, of course, matter for a lecture on function theory.

More interesting and easier to understand than this digression in function theory is the *geometric interpretation*. A decisive answer to the question of the geometric characteristic properties of algebraic curves has already been given, around the year

1850, by Hermann Graßmann[125]. Unfortunately, his interpretation has not yet been recorded in textbooks. I immediately announce Graßmann's theorem using the term "lineal *mechanism*".

A curve is algebraic if it can be generated by a lineal mechanism.

First of all, what is a lineal mechanism?

A *lineal mechanism* is a system of straight lines and points, partly movable and partly fixed, in which the movable straight lines are forced to pass trough certain (not necessarily fixed) points, and the points are forced to move on certain (nor necessarily fixed) straight lines.

For a better understanding, consider a certain number of straight lines and points as given. According to *Graßmann*, a *lineal construction* means the generation of new points and lines from the given fixed and movable ones with the sole help of a ruler[126]. The lineal construction becomes a straight-line mechanism if one requires that it leads back to the initial element.

I explain, for instance, how conic sections can be obtained with a Graßmannian mechanism:

We assign three points a, b, c and two lines B, Γ (Fig. 73). A point p moving on the plane abc describes a conic section if its motion satisfies the condition that the points β and γ lying on B and Γ, which are derived from p lineally (with the help of the lines $\overline{bp}$ and $\overline{cp}$), are in a line with the point a (*Colin Maclaurin's generation*).

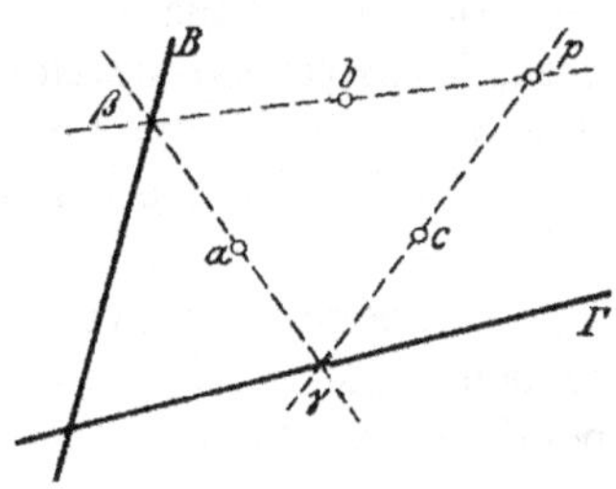

Figure 73

[133] With the use of rigid wires, or better of knitting needles, we can imagine the construction of an apparatus, so that the point p continues to satisfy the exposed lineal conditions, and moves along a conic section. This is then a lineal mechanism. In general, every mechanism of this kind has as the consequence that 3 points derived from the point p by lineal construction lie on a straight line, or also that 3 lines derived from the point p by lineal construction meet at a point. Lineal mechanisms have, by the way, only a theoretical interest, because the execution cannot be precise; indeed the apparatus trembles – however well one might design it.

[125] [See *Hermann Graßmann's* Gesammelte Werke, Vol. I, 1, pp. 245–248, Leipzig 1894, and Vol. II, 1 Abhandlungen 2–7, 14 and 18 (1904).]

[126] Translator's note: The German term for ruler is 'Lineal' – this explains Graßmann's terms.

Graßmann has used this idea of lineal mechanism to geometrically define an algebraic curve.

In the generation of a conic section, where the aim is to generate an algebraic curve of the second order, the point p is used twice. Graßmann finds out, that to generate an algebraic curve of order n, he has to set up the mechanism so that p is used n times[127].

He did not stop here; but he went on to show that, inversely, any algebraic curve $F(x, y) = 0$ of order n can be substituted by a (possibly very complicated) lineal mechanism, the construction of which applies the point p n times. He proves this in a way by which he even transforms the equation $F(x, y) = 0$ into a rule for the construction of such a mechanism.

After having established this theorem, Graßmann's purely geometric definition of an algebraic curve – given above – follows immediately: An algebraic curve of the order n is a curve that is generated by a lineal mechanism of order n.

We ask again, *what is the particularly interesting property in algebraic curves.*

Of course it also holds for such curves (as it does in general for analytic curves) that the total shape is determined by the behaviour on a very small piece. Furthermore, such curves have other *simple properties*, which concern their general development. To mention only the most important aspects:

They are made up of a finite number of branches, and the *theorem of Bézout* holds for them, namely that an algebraic curve of order n is cut by a straight line in n points, and that two different algebraic curves C_m and C_n intersect in $m \cdot n$ points (provided that we count correctly, that is we consider also the imaginary intersec- **[134]** tions, the points at infinity, and count every point with its multiplicity, disregarding the cases in which the curves have common branches).

On the basis of these general propositions, beautiful theorems arise, particularly for the curves from the first to the fourth order, which render it a real pleasure to work with these curves and even entail an aesthetic satisfaction.

For this reason, algebraic curves have been at times one of the favourite objects of mathematical studies. But to speak only of algebraic curves, as is done in many textbooks, must be considered as one-sidedness. There, one even gains the impression that only these exist, while the sine-curve or the helix occur much more often and are at least as beautiful curves, but not algebraic.

More special than the algebraic curves are the *rational curves*. Here one starts from the parametric representation $x = \varphi(t)$, $y = \psi(t)$, supposing that φ and ψ are rational in t. These, also, have been a main object of study.

---

[127] [The fact, that an algebraic curve of order n is characterized by a lineal mechanism should not be understood in the sense that every algebraic curve of order n can be constructed with the sole use of a ruler. It is not in general possible to determine with the sole use of a ruler a point p so that the incidence conditions of the lineal mechanism are satisfied.]

Mastery of the Empirical Phenomena by Means of Idealised Structures: Perry's Point of View

Having investigated the specialisations, which the concept of curve reaches in the field of precision mathematics, we ask – according to our general problematic:

What do the various idealised curves have to do with the applications?

According to an old prejudice, only analytic curves occur in applications. We argue contrary to this, as already often emphasised, that no idealised curves occur in the applications, but only approximations of them (e.g. stripes). If, nevertheless, one wants to consider an idealised curve so as to describe in an idealised way an empirical curve or a phenomenon, then it is sufficient to consider regular curves (that is, curves for which φ and ψ are differentiable a finite number of times, and the last pertinent derivatives might perhaps be at least piecewise monotonous). It is metaphysics to believe, instead, that one should use analytic curves. I call here metaphysics what is situated beyond the immediate experience, in this case: what is introduced only because one considers it desirable to use only analytic curves.

By the way, the assertion that analytic, algebraic, and rational curves per se have nothing to do with applications is somewhat mitigated by the fact that the simplest known regular curves are effectively analytic.

For instance, if we start with the approach

$$x = a_0 + a_1 t + a_2 t^2$$
$$y = b_0 + b_1 t + b_2 t^2$$

[135] (what one would always do, for the sake of simplicity, when one has to represent a particular piece of curve in the empirical field), then we meet by chance a rational curve. Analogously, in the applications, one will never abstain from the formulas $y = \sin x$, $y = e^x$, ..., even if additional corrections should come up that blur their analytic character. So we can add to the former considerations the following:

Analytic, algebraic, and rational curves occur in the empirical field only in so far as the simplest known regular curves always used for the approximated representation of empirical data are analytic, algebraic, and rational curves.

I want to add a few comments on the book: *John Perry*, The calculus for engineers[128], where the idea that I touch on here shows its advantage in a characteristic way.

The book is distinguished by the fact that its author, who writes for practitioners, takes on completely the point of view of the practitioner. He does not speak of the general concept of curve, etc.; instead, he begins with the fact that in practice actually only three fundamental functions ($y = x^n, y = e^x, y = \sin x$) prove to

[128] London 1897, translated into German by Robert Fricke and Fritz Süchting, Leipzig 1902. 4th ed. 1923. – [On Perry's movement see also Vol. II of this Elementary Mathematics, pp. 232–237, 280.]

be relevant, and that all tasks of the usual practice can be solved with sufficient exactness admitting only such functions that can be obtained from the three given ones by a finite number of applications of the four basic operations, for instance

$$y = \frac{a + bx + cx^2}{d + ex + fx^2} \text{ etc.}$$

This conception of *Perry* has often been claimed to be non-scientific. However this is not the case, considering its field. It is exactly the purpose of the present lecture course to choose the point of view in such a way that also Perry's ideas find their place in the edifice, which at the same time encloses the idealised and the empirical field.

II. Further Considerations of Precision Mathematics on Plane Geometry

Iterated Inversion with Respect to Two Touching Circles

I now take up again the ideas outlined before (thus turning completely to the ideal field). We have related our entire development to the ideas of analytic geometry and therefore, in the end, completely to the modern concept of number. If one wants to detach what has been discussed till now from this kind of representation and treat it at a purely geometric level, it is necessary to start with an axiom that corresponds to the introduction of the modern concept of number and that can be expressed as **[136]**
follows: *For every indefinitely decreasing sequence of closed intervals (segments, parts of curves, plane regions, parts of spaces), each of which contains all the following ones, there exists one and only one point common to all the intervals. This point is therefore univocally defined by the interval sequence.* This axiom, that we already happened to use, is called *axiom of the nested intervals*[129].

You can take this axiom as a theoretical basis for what I am going to say now. We have already seen that in the theory of automorphic functions one is led to define and to study point sets, which have all the remarkable properties treated in set theory.

Now, with a slight modification of the same geometric generation principle, I want to switch to *non-analytic curves*, which will be now generated purely geometrically – so that one can also avoid the impression that considering non-analytic curves introduces something artificial and alien into the field of geometry.

In our previous considerations we started from three or more completely disjoint circles on the plane, and reflected again and again in the circles their common confining region and all the new equivalent regions obtained by inversion (Fig. 74). So we obtained a net of regions that filled the plane except for an infinite set of limit points, and the following question arose: What can we say about this limit point set? We found out that it is nowhere dense, but nevertheless perfect and has the cardinality of the continuum.

[129] [It is easy to understand why the above formulation requires that the intervals be closed. The indefinitely decreasing sequence of open intervals $0 < x < \frac{1}{n}$, where n runs through the sequence of the natural numbers, does not contain any point that belongs to all the intervals. The point 0 is in none of the intervals.]

F. Klein, *Elementary Mathematics from a Higher Standpoint*,
DOI 10.1007/978-3-662-49439-4_8

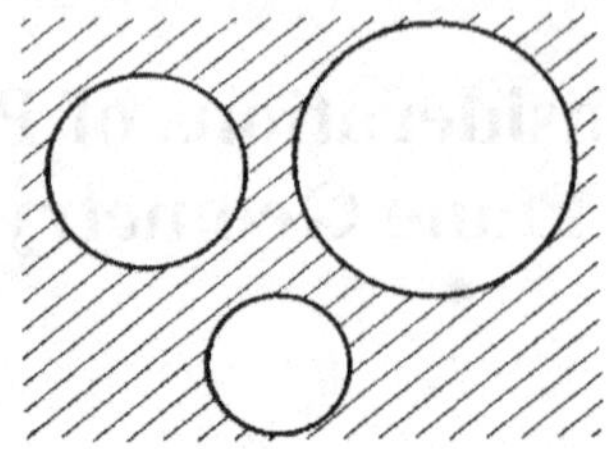

Figure 74

Now we introduce the modification, in that we allow the circles to touch each other in a cyclic succession, maintaining in any case the fact that there occurs no superposition of any two of the circular surfaces. In this manner the initial region is divided into two polygons with circular arcs and n edges, an inner and an outer one, each of which reveals but angles of magnitude 0. See Fig. 75, in which one of the
[137] regions goes to infinity, but this is of no particular relevance because an inversion immediately transforms it into a finite one.

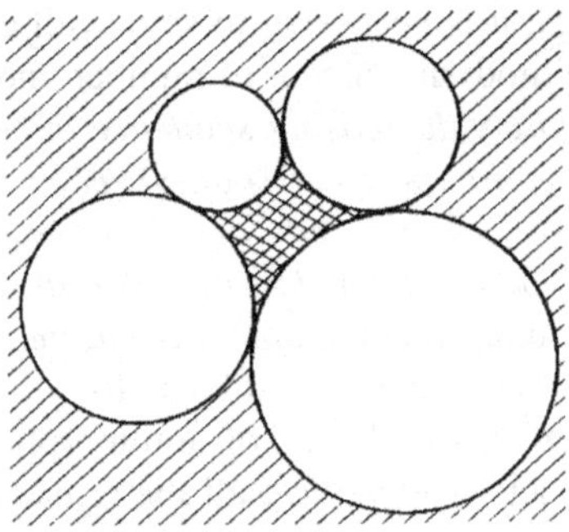

Figure 75

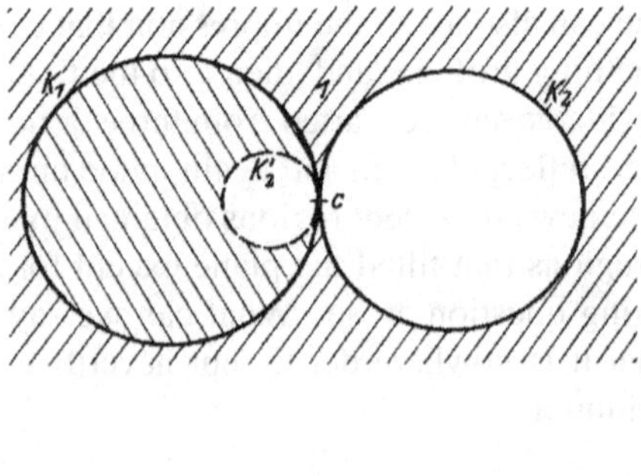

Figure 76

Now we apply unlimitedly to the zero-angle polygons with circular arcs the inversion with respect to the boundary circles and ask for the set of the limit points thus arising. *We will find that the set formed by the limit points and their accumulation points forms a closed Jordan curve, which is in general not analytic* – I will explain the qualification "in general" in the course of the discussion. This Jordan curve separates the net of regions formed by the external initial polygon and its images from the net of regions formed by the internal polygons and their images.

Here again I stress that the idea of mirroring again and again the initial regions and their equivalent regions with respect to a certain number of initial circles, which touch each other cyclically, has its origins in physics, since electrostatics induces one to observe exactly the same infinitely many inversions – we are now considering – in the equilibrium distribution of charges on rotating cylinders that are touching each other. If only one sufficiently deepens the considerations that anyway occupy practitioners, it is not possible to avoid non-analytic curves.

Let us first consider two circles K_1 and K_2, touching one another exteriorly; here, clearly, the contact is not cyclic. We thus have an initial region that can be described as a bi-angular circular arc polygon with two zero angles. It is denoted in Fig. 76 with 1, as before. If we reflect region 1 in K_1, then we obtain a sickle-shaped region enclosed by K_1, which we denote as S_1 like also the inversion considered; analogously, through reflection in K_2, we obtain a sickle-shaped region S_2 in K_2 (Fig. 77). Now we take again into account our afore-mentioned *principle of continuation*, which – instead of using the initial inversions as such – reflects again and again every newly obtained region in its inner boundary. It was proved before **[138]** that this principle coincides with the requirement to apply all combinations of the inversions S_1 and S_2 to the initial region. However at the same time we gave an *extended form* to the principle, obtaining that not only the confining region, but also all the regions constructed up to a certain moment, are reflected in each newly attained boundary circle. So, in our case, if we start with the three regions 1, S_1, S_2 (Fig. 77) we immediately obtain through reflection in K_2 three new regions, etc. Now, whether one wants to proceed in the one or the other way, anyway our simple case yields as result: *We obtain for our equivalent regions only one limit point, and this is identical to the point of contact of the two initial circles.*

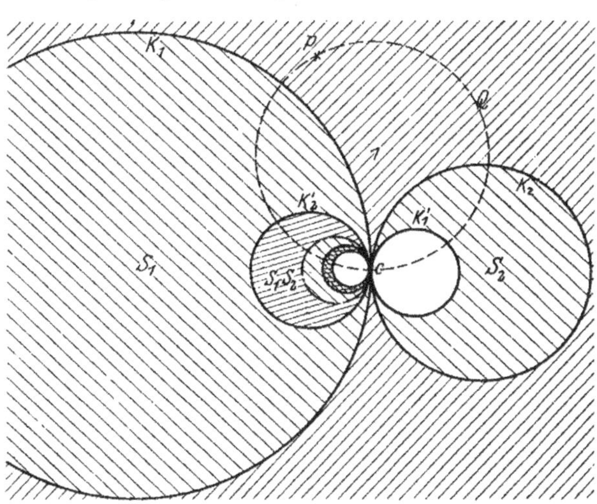

Figure 77

It is interesting to see here how, choosing an arbitrary point p in 1, the infinitely many points equivalent to p gradually approach the limit point. We now observe the *circle* Q[130]*, orthogonal* to the initial circles and passing through p (Fig. 77). We know that it is transformed into itself by all the operations $S_1, S_2, \ldots$ In fact, it passes through the point of contact c of the two circles K_1 and K_2; c and the tangent at c to Q are invariant with respect to all the $S_1, S_2, \ldots$; with respect to S_1 also the intersection point between Q and K_1 is invariant, as is also, with respect to S_2, the intersection point between Q and K_2. The consequence is therefore that all points obtained from p by inversion lie on the same orthogonal circle. We can even consider the orthogonal circle as a "trajectory" upon which p approaches the limit point by jumps.

This is of particular interest when we connect the limit point c through a straight line with an arbitrary point p of the plane. In fact, if we continuously apply to the point p the inversions corresponding to our principle of continuation, p moves toward c on its orthogonal circle, while each connecting line necessarily approaches a certain limit position, namely the "*central line*"[131] of the two initial circles (Fig. 78).

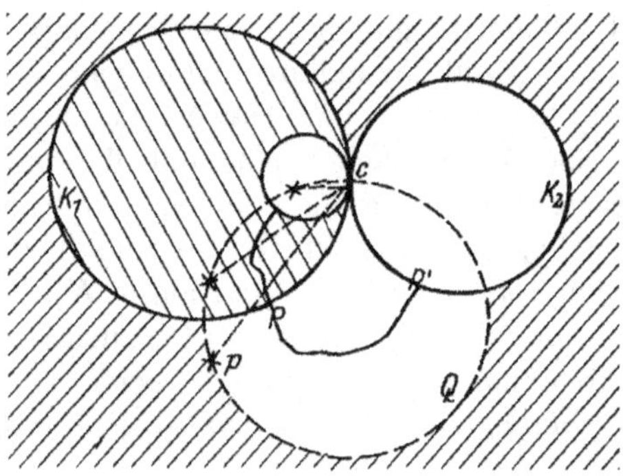

Figure 78

[139] Suppose, now, to consider a closed curve obtained by tracing an arbitrary piece of curve[132] (possibly non-analytic), which does not contain the point c and leads, in the original region, from the circle K_1 to the circle K_2; and that one continues to construct – thanks to our inversions – the totality of the equivalent pieces of curve, and at the end adds the point c, *then I maintain that the curve closed in this manner has a certain tangent at point c, namely the central line.*

---

[130] Translator's note: In the German version wrongly denoted as K.

[131] Translator's note: The term is equivalent to "normal line", as Klein will explain later on.

[132] [Of course, here is meant a piece of curve defined by equations $x = \varphi(t)$, $y = \psi(t)$; it is not difficult to prove that in the given way a closed curve (topologic image of a circle) arises.]

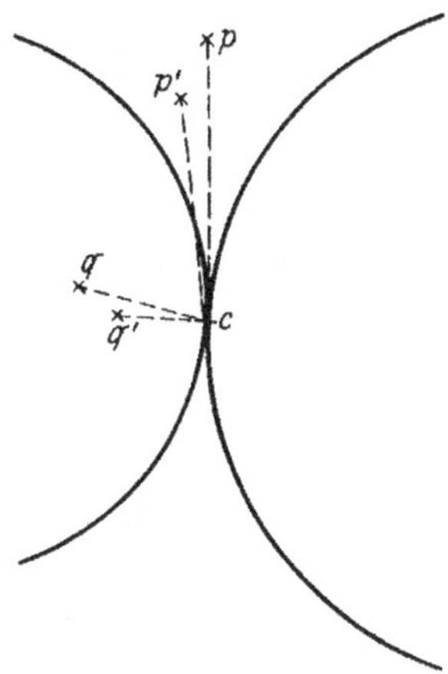

Figure 79

This theorem is a simple consequence of the auxiliary proposition just established[133]. Suppose we choose for the proof (Fig. 79) an arbitrary sequence of points $q, q', \ldots$ on the curve that converges to c. All the points $q, q', \ldots,$ which we will pass as we are approaching c, have points p, $p', \ldots$ equivalent to them in the initial region and it is possible to clamp the sequence of the $q, q', \ldots$ between the sequences P_ν resp. $P'_\nu, \ldots$ of the points equivalent to p resp. $p', \ldots$ (see Fig. 78). Consequently, the segments $\overline{cq}, \overline{cq'}, \ldots$ have the same limit position as the $\overline{cP_\nu}$ resp. $\overline{cP'_\nu}, \ldots$ but, according to the former auxiliary proposition, this is the central line.[134].

The Same with Respect to Three Touching Circles (Modular Figure)

Now I pass from this simple case of 2 initial circles to that of three initial circles. For convenience we choose them all of equal size. The initial region is now made up of two zero-angled curvilinear triangles (Fig. 80). Now, we know that three such circles always have *one and only one orthogonal circle Q*. This circle passes through the three contact points. It is invariant with respect to all the considered inversions, so that we can take it as a useful reference to get along in the figure.

---

[133] Translator's note: The one referring to Fig. 78, as clarified in the next note.

[134] Translator's note: This paragraph contains various typos, such as p, p' written in capital letters, or the omission of the three dots indicating that the sequences continue. We outline the proof contained in the first edition of 1902: "*The theorem is a simple consequence of the previous auxiliary proposition (which states that the line $\overline{cp}$ has, for any p, the central line as its limit position, when it is subjected to repeated inversions). Let us choose an arbitrary point q close to c, and trace $\overline{cq}$; we then proceed along the curve till we reach q', which is closer to c, and trace $\overline{cq'}$, etc. All the points $q, q', \ldots$ that we encounter when getting closer to c, have equivalent points p, p' in the initial region and arise from them through repeated inversions. Therefore the lines $\overline{cq}$, $\overline{cq'}, \ldots$ have the central line as their limit, thanks to the mentioned auxiliary property*"

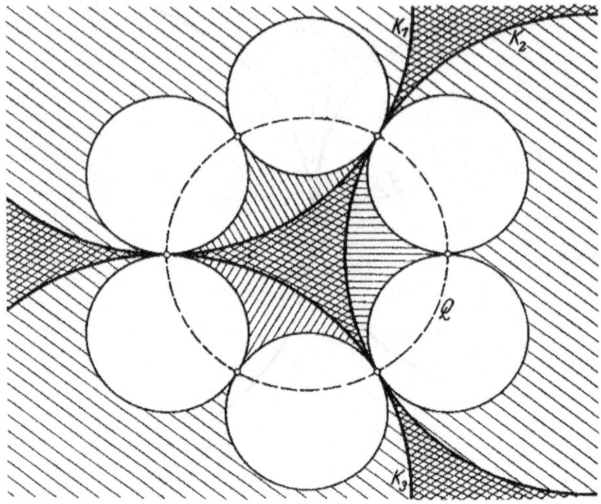

Figure 80

By the way, to obtain regions equivalent to both the initial regions, we need only
[140] to construct the equivalent regions internal to the orthogonal circle and then reflect in the orthogonal circle all the equivalent regions thus obtained. The correctness of this consideration follows from the proposition discussed earlier, namely that figures, which are inverse with respect to a circle K, yield – when they are inverted with respect to a new circle Q – figures that are inverse with respect to the circle K', arising when inverting K over Q. *Therefore, in our constructions, we limit ourselves to the internal part of Q.* Here we find at first three collateral triangles confining with the initial one; the four triangles together form a circular hexagon, whose tips are mounted perpendicularly on Q. According to our principle of continuation, we can now reflect either the 3 adjacent triangles separately in their free sides or also directly the whole hexagon in each of its 6 sides, obtaining, through this transformation, a 30-gon. You can see how we proceed:

The regions that arise from the initial region fill the interior of Q more and more completely, and the vertices of the obtained regions lie perpendicularly to Q in an increasing number. The further you run the process, the more densely the regions – equivalent to the initial region – approach Q, so that every point of the orthogonal circle becomes an accumulation point for region vertices, which we are going to show more in detail.

The figure described here, which is well-known among those who have dealt with elliptic modular functions, is in a certain sense a gift that geometry has received from the theory of functions. Because even though such an infinite process can very well be initially considered from a purely geometric standpoint, yet the historical development has been that it was within the *theory of the elliptic modular functions* that one was obliged to study such figures. From a purely conceptual point of view this figure can be already found in *von Staudt* (in that he constructed again and again the fourth harmonic point given 3 arbitrary points of a conic section), but *von Staudt* has not developed this from an intuitive point of view. [141]

I take a closer look at our figure with the three touching circles by highlighting some details, which make the study of the figure in the case of four touching circles much easier.

I constructed the figure symmetrically choosing equal radii for the touching circles, so that their points of contact coincide at the vertices of a triangle inscribed in the orthogonal circle Q. We obtain the corresponding complete figure either by first multiplying the internal region by inversion and reflecting the partial figure thus obtained in the orthogonal circle Q, or by first multiplying the external region and then reflecting the partial figure thus obtained in the orthogonal circle.

We now want to observe how the points of contact of the arising curvilinear triangles accumulate on the orthogonal circle. We naturally have 3, then 6, then 12 etc. points of contact, so we can say: *Proceeding with the construction, the orthogonal circle is filled more and more densely with the points of contact of two circles.*

The next question is then: What can be said about the set of points of contact? *It is easy to see that in the set of the points of contact each point is an accumulation point for the others, that is, the set is dense in itself on the orthogonal circle.*

To prove this theorem, we consider any two touching circles, “1”, “2”, obtained during the construction process; we call c their contact point, which lies on the orthogonal circle Q. Now we imagine this figure (Fig. 81) transformed as usual through an indefinite sequence of reflections. First we reflect the figure in the two circles “1”, “2”, then the thus completed figure in the new boundary circles, etc. Consequently , the point c, which was a point of contact of only two circles before the first reflection, becomes a point of contact of infinitely many circles. [142]

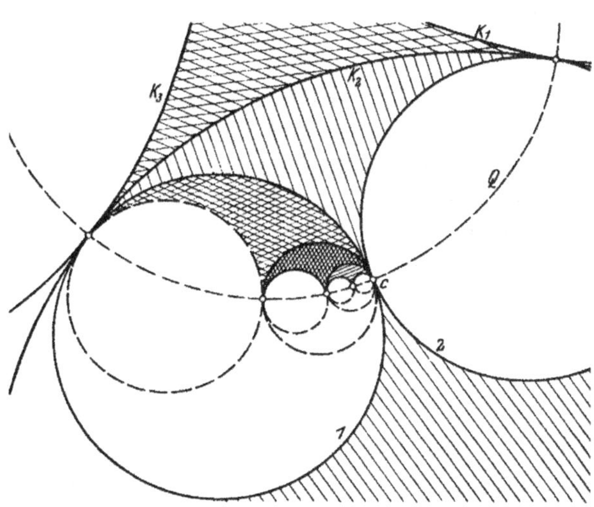

Figure 81

And each of these infinitely many circles, which become arbitrarily small as the inversion process is continued, presents a new contact point at its second intersection with the orthogonal circle. Figure 81 shows how these additionally arising new points of contact all belong to curvilinear triangles lying side by side. Figure 82 shows the particular case in which the orthogonal circle degenerates into a straight line.

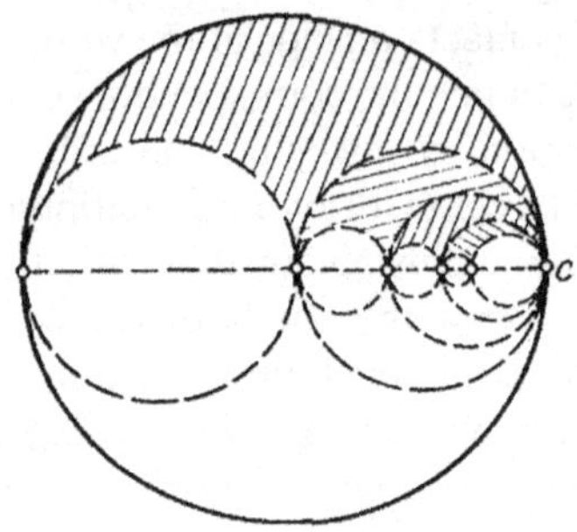

Figure 82

Every circular neighbourhood of c, however small, therefore contains infinitely many points of contact and c is an accumulation point for the contact points. But c was a completely arbitrary contact point. Consequently, the set of the points of contact is dense in itself.

Among the other properties of this point set I mention that *it is everywhere dense on the periphery of the orthogonal circle*, that is, on every part of the periphery of the circle, however small, lie contact points. This can be seen as follows: our initial figure (Fig. 80) forms a closed sequence of three circles, which are aligned along the orthogonal circle. If we reflect in these three circles, we obtain a new closed sequence of six circles that again lie along the orthogonal circle. In general, every reflection in the totality of the circles of such a sequence yields a new sequence with more numerous and smaller circles. Now we consider an arc on the periphery of the orthogonal circle, corresponding to a chord of length δ. We then push the process of reflecting so far as to obtain a sequence of circles, whose diameters are all smaller than δ. Then, for any δ, however small, there surely lies on the arc one contact point, and it is easy to see that – continuing the process of reflection indefinitely – infinitely many points of contact will lie on the arc. It follows at the same time that other points are added to the set of contact points, namely all those points on the orthogonal circle that are not points of contact but accumulation points of the set – that such points exist will soon become clear.

The relation between the everywhere dense set of points of contact on the orthogonal circle Q and the remaining points of Q is similar to the relation between rational and irrational points on the x-axis. On this axis, as well, the "rational"
[143] points lie everywhere dense and are denumerable, while the "irrational" points appear as accumulation points for the rational numbers.

If we consider that, according to Dedekind, irrational numbers are defined by means of a cut in the set of rational numbers, endowed with certain qualities, the analogy discussed becomes even clearer: Every point of the orthogonal circle, which is not a point of contact, effects a cut in the set of points of contact and can even be defined via such a cut.

Until now I have spoken of these things in a somewhat indeterminate manner, because I did not refer to any quantitative relations but only to the figure as such. However, it is easy to give to the figure a form that allows an arithmetic interpretation of everything.

In fact, we can give a simple form to the initial region – without a loss of generality, as we consider relations that are not changed by any inversion of the total figure. Let us reflect the initial triangle in a circle with its centre on a vertex of the triangle[135]. We thus obtain a triangle delimited by two parallel straight lines and by a semi-circle touching these lines (Fig. 83). The orthogonal circle is stretched into the line QQ, which passes through the two finite contact points. Upon this figure usual considerations can be made, as they are performed in function theory (where the figure is also placed in a convenient coordinate system).

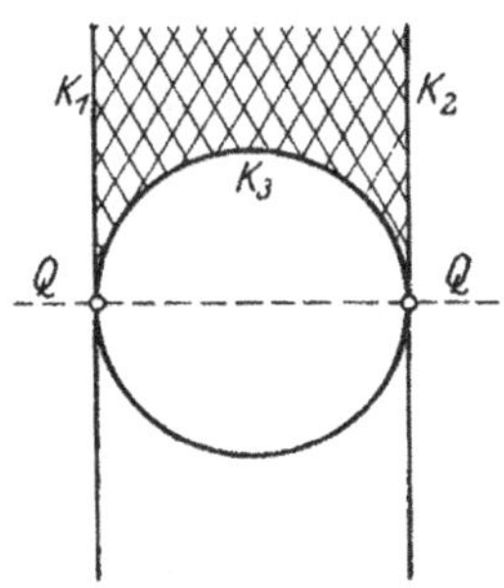

Figure 83

We choose the line QQ as x-axis, and the origin and the scale so that the points $x = 0, x = 1$ fall at the two finite vertices. Then the whole figure is easy to construct, repeating the first figure unlimitedly on the right and on the left, and reflecting the sequence of infinitely many triangles such obtained in each of the occurring semi-circles (Fig. 84). In this way all the points of contact can be found on the x-axis, and it is easy to calculate exactly arithmetically what had been explained in our first figure only with the help of our immediate geometric sense. The for- **[144]**
mulas occurring then indicate that all the points of contact have rational abscissas x and that every point with a rational x becomes a contact point. *So, there is not only an analogy between the set of the points of contact and the set of the rational points on the x-axis, but there exists an identity. In particular it results that the set of the points of contact is denumerable.*

This serves as complement to the case of three circles!

---

[135] Translator's note: The orthogonal circle degenerates to a line. One of the half-planes is seen as internal to the orthogonal circle.

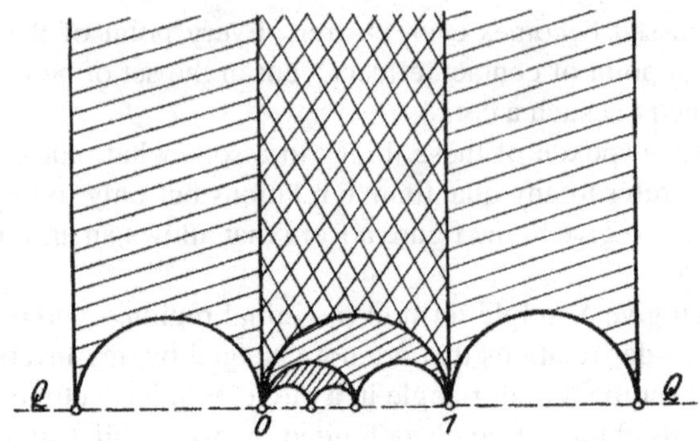

Figure 84

The Normal Case of Four Circles Touching Each Other in a Cyclic Succession

Now I switch to our proper subject, where we start with 4 circles touching each other in a cyclic succession. The situation here is similar to the previous case.

The 4 circles may at first still have a common orthogonal circle (Fig. 85). In this case we can take as initial region the outer or the inner part of the two arising
[145] curvilinear quadrilaterals[136]. If we take the inner part and use the inversions to construct the adjacent quadrilateral we obtain first of all – due to our principle of continuation – a dodecagon made of 5 quadrilaterals, then a 36-gon etc. The number of points of contact on the orthogonal circle becomes, with this highly symmetric procedure, 4, 12, 36, 108, ... We immediately ask how the points of contact are distributed along the periphery of the orthogonal circle. We can state the following:

1. *The set of points of contact is denumerable*
2. *The set of points of contact lies everywhere dense on the orthogonal circle.*

We now want to further specialise our figure in that *we bring, with a suitable inversion, a point of contact to infinity*. The result is Fig. 86, and this figure must be conveniently chosen as a basis for the arithmetical treatment.

The orthogonal circle degenerates into a straight line that we take as x-axis, two of the circles again become parallel lines, while the two other circles between these two lines lie orthogonal to the x-axis. The complete figure arises through repeated reflections to the right and to the left in the lines and the circles. If we place two points of contact in 0 and 1, and denote as σ the abscissa of the third point lying in the finite domain ($0 < \sigma < 1$) we obtain a set of contact points on the x-axis, which

[136] We define external curvilinear quadrilateral the region that contains the point at infinity.

no longer coincides per se with the set of rational points, but which has abscissas that depend *rationally on the quantity* σ. It is interesting to investigate the qualities of these rational functions of σ.

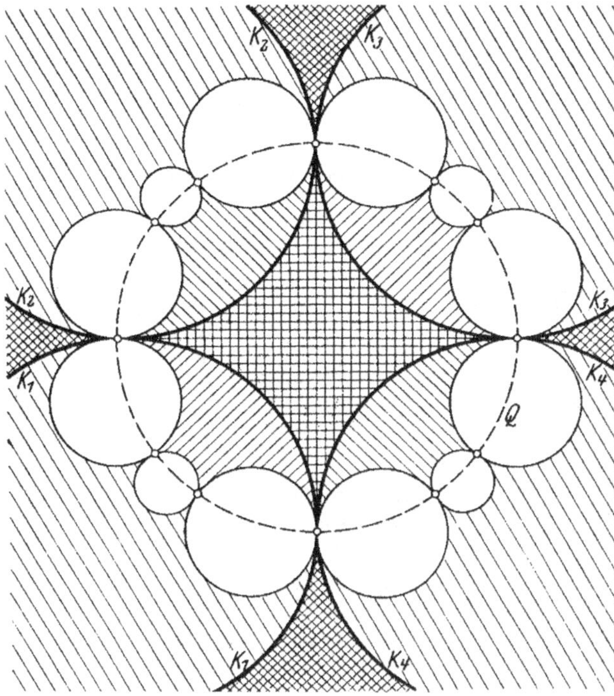

Figure 85

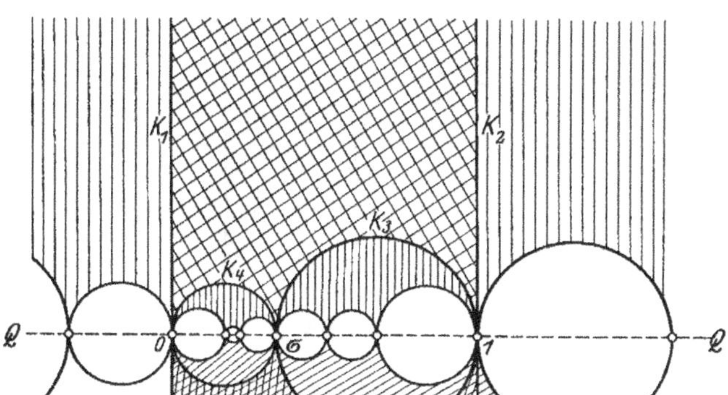

Figure 86

The General Case of Four Circles Touching Each Other in a Cyclic Succession

This was the particular case, in which there exists a common orthogonal circle for the 4 initial circles. *Now we pass to the obviously more complicated general*
[146] *case.* Thanks to the theorems achieved before, we will also gain some overview of this case. Let four circles be given, which touch one another in a cyclical succession. Then it is remarkable that the theorem still holds according to which the four points of contact *lie on a circle*, which is however generally not an orthogonal circle (Fig. 87).

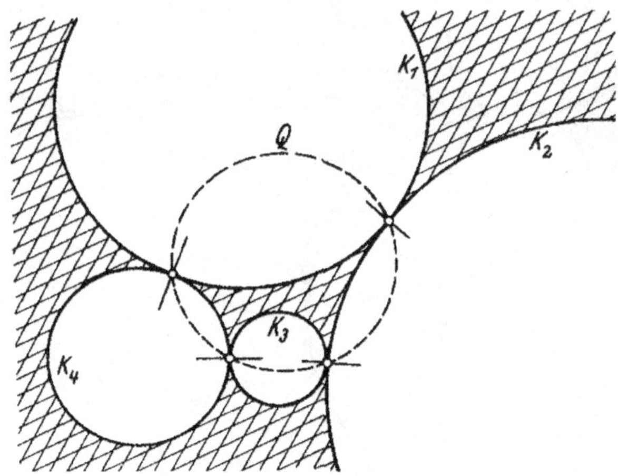

Figure 87

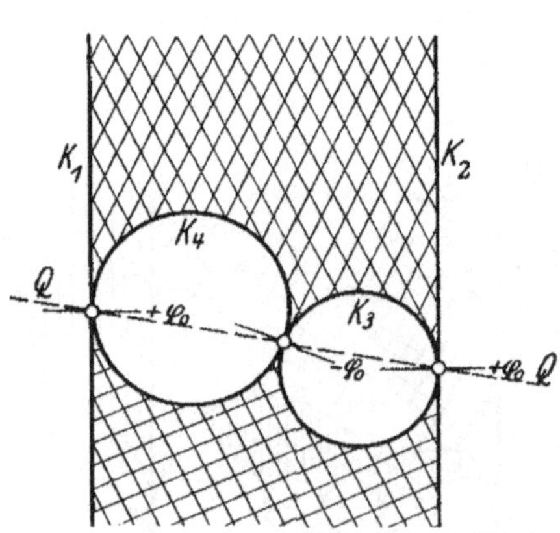

Figure 88

To prove this auxiliary theorem we apply the artifice already used, simplifying the figure by means of an inversion in a circle that has one of the points of contact as its centre. The two circles, which are touching each other with angle zero at this centre of inversion, are transformed through the inversion into two parallel lines K_1 and K_2, between which lie the two other circles K_3 and K_4, as shown in Fig. 88.

If we want to show that in our original figure the four points of contact lie on a circle, then we need to prove for Fig. 88 that it is possible to find a line passing through the three contact points belonging to the finite domain. But this is immediately clear from the point of view of elementary geometry, and our auxiliary theorem is thus proved. We call this line the *transverse line* Q, and generally the circle through the four vertices of the curvilinear quadrilateral the *transverse circle* Q.

At the same time it is also clear that, drawing in Fig. 88 the normal lines to the circles at the points of contact (which intersect each other at the centres of K_3 and K_4), our transverse line cuts these normal lines or *centrals*, as I prefer to say more clearly, forming two equal oppositely oriented non-zero angles $\pm\varphi_0$; the meaning of the signs is clear from Fig. 88. If it were $\varphi_0 = 0$ we would be back in our previous case. By the way, φ_0 is, as we can easily prove, always less than $\frac{\pi}{4}$. If we apply this situation to our original figure, we can say: The four points of contact lie on the transversal circle Q, which forms with the centrals passing through the points of contact angles $\pm\varphi_0$, where φ_0 is different from zero, because Q is no longer an orthogonal circle, but smaller than $\frac{\pi}{4}$. This theorem can serve to give to the initial [147]
figure a form that is convenient for further considerations. It is well known that an intricate figure becomes more intelligible if it is presented in a symmetric form. So, in the following, the initial figure may be drawn so that we start with the transversal circle Q and choose the four points of contact at the vertices of a rectangle inscribed in the transversal circle (Fig. 89).

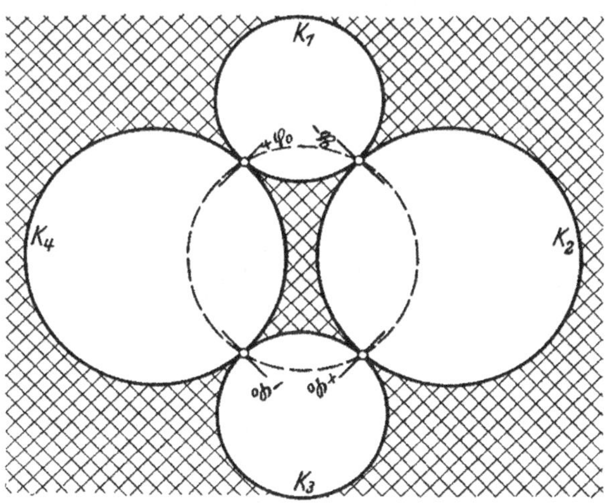

Figure 89

If, after a suitable choice of an angle $\pm\varphi_0$ $\left(\varphi_0 < \frac{\pi}{4}\right)$, different from zero, we draw the slopes of the central lines, it is easy to construct the 4 "initial circles" afterwards. We thus obtain 2 curvilinear quadrilaterals, an inner and an outer one, which are not inverse with respect to the transversal circle Q, but which maintain the same relative position as in the case of inverse figures. Now we reflect, at first,

Q in K_1 obtaining a new circle Q'. The construction of Q' becomes simplified by the fact that the intersection points of K_1 with Q are fixed and the angles $+\varphi_0$ and $-\varphi_0$ maintain their size, though they change their sign. The consequence is that the arc q' of Q', which is the only one interesting to us and which falls into the interior of K_1, turns its concavity toward the interior curvilinear quadrilateral. The reflection of Q in K_2, K_3, K_4 is likewise easy to construct. We obtain the circles Q'', Q''', Q^{IV}; we may call q'', q''', q^{IV} their arcs falling inside K_2 resp. K_3, K_4. The arc q''' turns, as does q', its concavity toward the internal curvilinear quadrilateral, while q'' and q^{IV} turn in this direction their convexity. Therefore, after the four inversions, the transversal circle Q is substituted by the four arcs q', q'', q''', q^{IV}. These four arcs form angles $\pm\varphi_0$ with the centrals of the points of contact lying on them, and, as can be easily seen, are lined up at the points of contact without forming corners (Fig. 90).

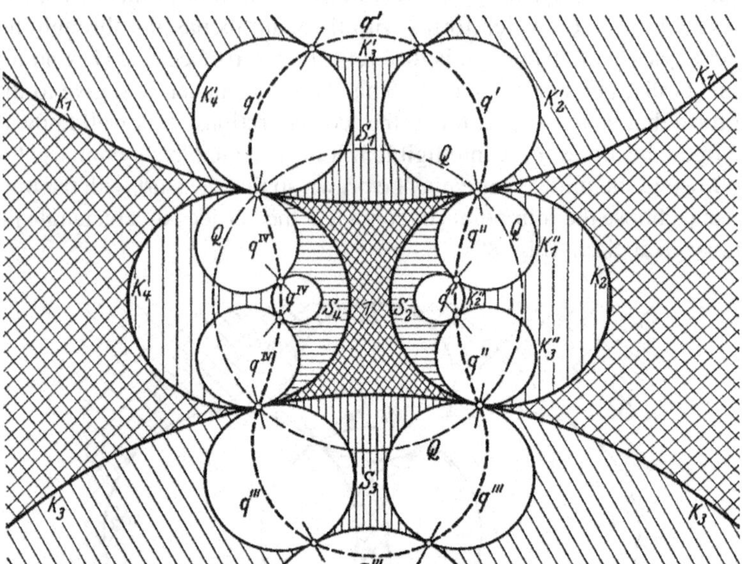

Figure 90

After drawing this new transversal curve, we reflect the circles K_2, K_3, K_4 in K_1. The new points of contact will lie once on the arc q', and once on the straight line
[148] passing through the centre of K_1 and the old contact points.

When we then draw the centrals of the new points of contact in the correct position, the images K_2', K_3', K_4' of K_2, K_3, K_4 will be easily found. K_2', K_3', K_4' delimit, together with K_1, a first lateral image of our two initial quadrilaterals. In complete analogy, the inversions can be performed of K_1, K_3, K_4 in K_2, of K_1, K_2, K_4 in K_3, and of K_1, K_2, K_3 in K_4. Since the transversal circle Q and the arcs q', q'', q''', q^{IV} are the simplest elements for easily getting a grasp on the figure arising, I want to describe the figure with the following words again:

At the beginning we had a circle with 4 points of contact and their centrals. From this circle Q we obtained a curvilinear chain made of 4 arcs, which has now in total 12 points of contact with 12 centrals. On the segments of the chain thus obtained 12 circles $K^I, K^{II}, \ldots, K^{IV}, \ldots$ are soaring. These divide the plane in an inner and an outer region.

It is clear how this construction should be continued. We single out one of the new circles, for instance K_2'. We reflect the two curvilinear quadrilaterals confining with K_2' in its interior. Thus we obtain three new circles in K_2'. The points of contact of these circles lie on that arc (q'), which arises in the interior of K_2' by inversion of q' over K_2'. The delimitation of the quadrilaterals we have just considered is shared, among the new circles, by K_3' and K_4', as well as by K_2' . Now if we reflect **[149]**
the quadrilaterals in the interior of the last two circles, we will also obtain in each of these three new circles and for each one an arc, on which the points of contact lie. The arc q' of Q' is substituted, after the three reflections, by a line made of three arcs. The three arcs are again lined up without making a corner, and form with the corresponding centrals alternately the angles $\pm\varphi_0$. Now let us look at the two curvilinear quadrilaterals that have been generated in K_2'. If we reflect both, together with the arc (q') of the contact points, in the three circles lying in K_2', then in each of these arise again three circles, two quadrilaterals and an arc for the contact points. At the end, the arc of the previous points of contact is again substituted by a chain of three new ones, which touch each other without making corners and form with the centrals of the touching circles alternately the angles $\pm\varphi_0$. This chain is shown in Fig. 91.

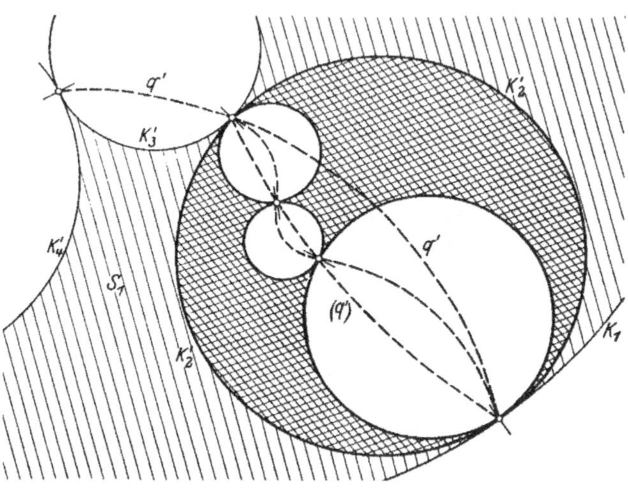

Figure 91

The substitution of an arc of the transversal circle with three new ones reminds us of the way in which we constructed the parametric curves $\varphi_{n+1}(t)$ resp. $\psi_{n+1}(t)$ from $\varphi_n(t)$ resp. $\psi_n(t)$ when discussing the Peano curve (Fig. 92). Let us now go back to the complete figure that we obtained when we reflected the two initial quadrilaterals over K_1, K_2, K_3 and K_4. We obtained $4 \cdot 3 = 12$ new circles, the 4 old points of contact and 8 new ones were all lined up on a closed transversal curve, formed by the arcs q', q'', q''', q^{IV}.

Figure 92

[150] If for every triple of the new circles we apply the mentioned reflections we obtain $4 \cdot 3^2$ new delimiting circles – ordered in a closed chain – and to the previously existing 12 points of contact twice as many are added. The 36 points of contact that we have now lie on a closed curve, which is composed by $4 \cdot 3$ arcs, lined up without making corners, and that forms with the central line alternately the angles $\pm\varphi_0$. If we apply the reflections n consecutive times, we obtain:

1. $4 \cdot 3^{n+1}$ new delimiting circles, through which the new transversal line passes, in the step from the n-th to the $(n+1)$-th reflection;
2. In total, $4 \cdot 3^{n+1}$ contact points;
3. $4 \cdot 3^n$ arcs, which form the new transversal line.

You understand the point. If we call the single transversal lines C_0, C_1, C_2 etc., the question is: Which is the limit form C_∞ of this sequence of transversal lines? Is C_∞ the locus of all points of contact and their accumulation points, the point set that divides the net of the internal quadrilaterals from the outer ones?

Properties of the Non-Analytic Curves thus Arising

If we already call C_∞ a curve, we anticipate what we still have to prove. However, we will show this designation to be justified; we will even prove the following theorems

1. C_∞ is a *Jordan curve*, that is, C_∞ can be expressed by formulas of the kind $x = \varphi(t), y = \psi(t)$, where φ, ψ have the known properties.
2. In all the infinitely many contact points, lying everywhere dense on it, it has *the corresponding centrals as tangents.*
3. *The curve is, however, not analytic at all*, as long as φ_0 is different from zero, what we assumed explicitly.

I first prove the second of these statements.

I prove the theorem relating to previous considerations that I now deepen in reference to our special situation.

We choose two circles that touch each other at the point of contact c – that we want to consider – and divide the plane in an outer and an inner region (Fig. 93).

Now we construct the whole point set C_∞ by drawing – or imagining as drawn – what will result situated outside our two contact circles (this is a point set, of which

we do not yet know whether it is a curve in the defined sense; the development that we are showing here is independent from this question) and by multiplying the [151]
figure obtained through repeated inversions in the interior of the circles K_1 and K_2; in this way the point of contact c becomes a limit point. The question is: Can we speak of a tangent to the set at point c?

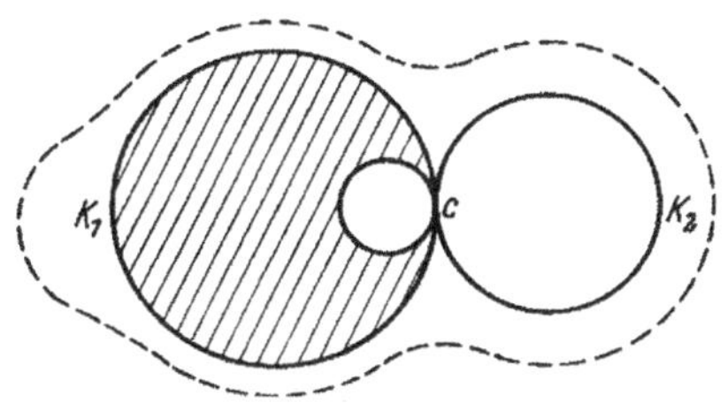

Figure 93

We will do this when a secant qc, where q is an arbitrary point of the set, tends to a certain limit position as q approaches c indefinitely, independently from the way in which the approach of q to c is chosen.

We say now: The point q might be chosen as one likes, there will always exist on the point set C_∞, in the external region, somewhere, a point p, which is transformed into q by inversion of S_1 in K_1, or by inversion of S_2 in K_2, or by a chain of these two inversions, and which is therefore equivalent to q. Consequently, to every secant $\overline{qc}$ corresponds a secant $\overline{pc}$. Therefore, it is possible to obtain all secants $\overline{qc}$ by drawing all secants $\overline{pc}$ and repeatedly applying the transformations S_1 and S_2 to the points p. Instead of asking for the limit position of the secant $\overline{pc}$ when q – always belonging to the set C_∞ – approaches c indefinitely, we first want to ask for the limit position of an arbitrary secant $\overline{pc}$, when the transformations S_1 and S_2 are indefinitely applied to p.

However, in this way we come back to our observations on p. [138]. There we saw that the limit procedure we are considering now yields for the secant $\overline{pc}$ the central line as its limit. This holds for every secant $\overline{pc}$; *all* secants $\overline{pc}$ have the central line as their limit form. Now we can show that also for every sequence of points q_n of C_∞ approaching c the secant $\overline{q_n c}$ is transformed into the central line[137]. Consequently, the central line has to be considered the tangent of the point set C_∞.

---

[137] [For every point of the sequence q_n there is an orthogonal circle of K_1 and K_2, on which it lies with all its equivalent points. Now, in the set of all orthogonal circles of K_1 and K_2 passing through c and on which lie points of C_∞, there are two extreme ones H_1 and H_2. The first of these two passes through the point of contact of K_1 and K_4, the second through the point of contact of K_2 and K_3. In fact, if we observe (see Fig. 87) the quadrilateral delimited by K_1, K_2, K_3, K_4, we see that we can obtain all the points of contact of C_∞ lying in these circles, if we apply to the vertices of the quadrilaterals indefinitely the transformations S_1, S_2, S_3, S_4. Images of a vertex can thus appear in K_1 resp. K_2 only on H_1 resp. H_2, or between them. The same also holds for the accumulation points of the contact points. Therefore, the curve C_∞ must squeeze through H_1 and H_2 as it passes through K_1 and K_2. So, the arbitrary sequence of secants $\overline{q_n c}(q_n \to c)$ can be clamped between two sequences $\overline{p_n c}$ and $\overline{p'_n c}$, where the p_n resp. p'_n are points that are equivalent to each other that lie on H_1 resp. H_2.]

This result is appealing if we think of the sequence of curves $C_1, C_2 \dots$ In fact,
[152] these form at the point c alternately the angle $\pm\varphi_0$ with the central line. But the limit curve C_∞ touches, as we now know, the central line itself.

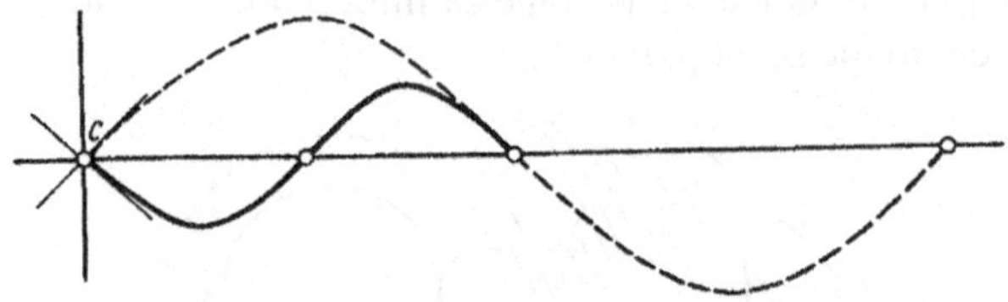

Figure 94

To make such a behaviour more plausible I construct the following example, which presents an analogous situation. As suggested by Fig. 94, we put through c a sequence of sine curves with amplitude $\frac{1}{2n}$, wavelength $\frac{\lambda}{2n}$, and phase $n\pi$ at $c(n = 0, 1, 2, \dots)$.

The tangents at c to these sine curves, which share with the curves $C_1, C_2 \dots$ the wave-form, will form with the x-axis alternately the angles $\pm\varphi_0$. The question is: Which is the tangent of the limit curve? The question obviously requires carefully observing the correct order of the limit passages. It is not the tangents of the single auxiliary curves $C_1, C_2, \dots$ that tend to a limit (they always jump), but the same curves have themselves a limit curve, namely the x-axis, which has a specific tangent at point c, evidently the same x-axis.

We now know that our set C_∞ of the points of contact c and their accumulation points has a specific tangent at every point. *But this does not mean that C_∞ is an analytic curve.* We will understand this in the following manner:

Suppose that two circles touching each other in c are taken from the sequence of circles appearing in our figure. In one of them we construct two quadrilaterals (Fig. 95). We call the four points of contact arising in this manner as shown in the figure: c, c^I, c^{II}, c^{III}. As we know, they lie on a transversal circle, which forms with the central in c the angle φ_0. The sign of the angle may be positive.

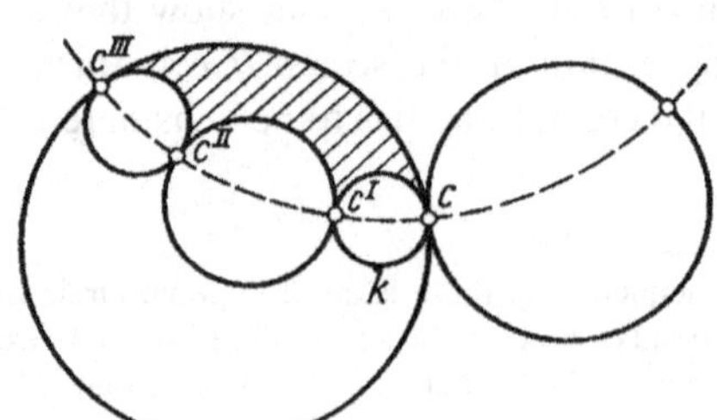

Figure 95

Not considering the point c^{III}, we have a circle passing through the three points c, c^I, c^{II}, which is inclined, with respect to the central line in c, by an angle $+\varphi_0$. We now reflect the two quadrilaterals and their orthogonal circle arcs in the circle

denoted in our figure by k. Then we reflect the quadrilaterals created in k, together with their orthogonal circle, in that circle of the three new limiting circles lying in k that passes through c. We consider this process of reflection as continuing indef- **[153]**
initely. The points c^I, c^{II} then move unlimitedly toward c. We obtain a sequence of points c^I, c^{II} all belonging, with c, to the point set C_∞. Every triple of points c, c^I, c^{II} lies on a circle, which forms with the fixed central of c an angle of constant absolute value φ_0. But this angle is alternately positive and negative. So we have for the point set C_∞ the following property, which cannot occur in an analytic curve:

It is possible to find two other points c^I, c^{II} in an arbitrary neighbourhood of a given point c, by infinitely many manners, so that the circle through c, c^I, c^{II} forms with the tangent in c an angle of $\pm\varphi_0$.

In fact, this is not possible in an analytic curve. Because, if we connect three points c, c^I, c^{II} of an analytic curve through a circle, and we let c^I, c^{II} approach c indefinitely, then we obtain a specific limit form, namely the circle of curvature, which, as such, touches the tangent in c. It is entirely impossible, that in an analytic curve our circle jumps back and forth forming with the tangent a finite angle $\pm\varphi_0$. *The curve C_∞, if ever it is a curve, cannot have a circle of curvature in any point c.*

It remains to prove that our point set is a Jordan curve.

According to the considerations on p. [123] we can call the point set C_∞ a closed Jordan curve if it can be transformed univocally and continuously from both sides into the points of the periphery. This can be done with our set C_∞ simply by comparing Fig. 85 with Fig. 90.

In fact, we can simply put in relation the four vertices of the initial quadrilateral of the one figure with the four vertices of the initial quadrilateral of the other figure, in a cyclic order, and then the vertices of the adjacent quadrilaterals of the first figure with the corresponding vertices of the adjacent quadrilaterals of the second. Thanks to the law thus expressed, the points of contact of the one figure are clearly put in relation univocally and continuously from both sides with the points of contact of the second figure. As the set of the points of contact lies everywhere dense in the complete figure considered, this correspondence can be extended to a likewise invertible univocal and continuous correspondence between the complete figures, by considering each point α of the orthogonal circle (Fig. 85) – which is not a point of contact – as limit point of a sequence of points of contact α_n and putting it in relation with the limit of the sequence formed by the images of the α_n. The point set C_∞ of Fig. 90 is therefore put univocally and continuously in relation with the points of a circle, and the proof is concluded. *So, the point set C_∞ is in fact a Jordan* **[154]**
curve[138].

---

[138] [This kind of non-analytic curves was first explored by *Henri Poincaré* (*Acta Mathematica* Vol. 3 (1883), pp. 77–80 = Œuvres, Vol. 2, pp. 285–287). *Poincaré*'s attention to the existence of these curves had been called by a letter of *Klein*. This letter of Klein to Poincaré is printed on pp. 590–593 of the third volume of Klein's complete mathematical works. See also, on p. 582 of the same volume, *Klein*'s observations on the prehistory of automorphic functions. In the particular case described above of the zero-angle curvilinear quadrilateral without orthogonal circle the non-analytic limit form arising through reflection was studied more in detail by Robert Fricke (Die Kreisbogen-

With the discussion we have just concluded of our automorphic figure I have in some way reached the apex of the precision geometry part of this lecture, and it seems therefore appropriate to take from this point on an overview on the results obtained.

By relating to the theory of automorphic functions, I succeeded in constructing – by means of a purely geometric process and excluding any arithmetic operation – at first a point set that is *nowhere* dense but perfect, and then, selecting particular situations in the initial figure[139], I produced a *non-analytic* curve. It became clear, thus, that these modern ideas emerge also on a purely geometric basis, and that they are not a simple fiction of the analysts. I express my general conviction in the following way:

The modern issues of set theory emerge everywhere, when precision mathematics questions are sufficiently deepened. One cannot avoid facing them.

So, for geometry (and for all other mathematical disciplines that have to do with space intuition) a two-part program follows, which is, however, mostly presented only one-sidedly or whose structure is not clearly understood:

1. *to consider and to cultivate approximation mathematics as such,*

2. *on the other hand, however, to not shy away from any idealization (in the sense of precision mathematics).*

I would say: One should "do one thing without leaving out the other".

I want to explain how *this conception* shows its relevance in *theoretical mechanics,* the science that most uses, together with geometry, space intuition.

[155] Here we have the questions:

What is, in theoretical mechanics, approximation mathematics, what is precision mathematics, to what extent are the two aspects developed as consequences of one another, to what extent they are mixed in an unclear manner?

We have this distinction between two separate situations:

a) There is a first kind of *representation of mechanics*, which only refers to observations and does not go beyond the limits of empirical data. It does not work on mass points but with small particles, etc. The tendency to describe, in the natural sciences, only what can be observed directly, without building theories on what may be beyond the appearance, has been called "*phenomenology*". By using this expression it becomes clear that *phenomenological mechanics* is a field that applies approximation mathematics. How this has to be understood in detail, emerges from our previous considerations. (There will be no reference to rigorous limit concepts; the mathematical description of the phenomena will be limited to the linear terms, because one can gain herewith a sufficiently exact approximation, etc.)

b) Next to that we have *idealistic mechanics*, as I call it. It develops mathematical representations that go beyond perception, and studies how to cope with them in a reasonable way. Here we have real *mass points*, rigorous laws, according to which these points interact, real derivatives, etc. This idealized mechanics belongs to the

vierseite und das Princip der Symmetrie, *Mathematische Annalen*, Vol. 44 (1894), pp. 565–599; see also *Fricke* and *Klein*, Theorie der automorphen Funktionen Vol.1, pp. 415–428.)]

[139] Where I supposed that the four initial circles touched each other in cyclic order.

field of precision mathematics. Here we must not shy away from any idealization, one should consider very general point sets, non-analytic curves, etc.

Instead of such a distinction, a different mathematical treatment is usually applied in mechanics, which is a mixture of a) and b). This treatment is customary, because the distinction between approximation and precision mathematics had not clearly emerged in the 18th century and also in the 19th century it has only gradually taken over. Therefore all functions are considered eo ipso as analytic, and all limit passages are considered as interchangeable ("as it was in paradise"; *Paul du Bois-Reymond*).

Now, I have always thought that a lecture course for beginners has to be held in this somewhat illogical way, however I stress that, as soon as a more mature understanding has been acquired, the second treatment, that is, the conscious separation of precision and approximation mathematics even in problems of mechanics, cannot be omitted. **[156]**

Premises to the Whole Development. Further Idealization by Veronese

If I may refer, in this context, to the treatment of mechanics in the *"Enzyklopädie der mathematischen Wissenschaften"*, then it is clear that there the modern conception (which I sustain) can only marginally become the guideline. This is because the collaborators must refer to the existing literature, as it is, and cannot force it into the frame of a new and often unfamiliar conception; indeed, the aim of the *Enzyklopädie* is only to be a *collection of the material*, so that further research is not limited, as it has often been, by a lack of knowledge of the existing.

But I also must recall, in this context, that there are mathematicians who *want to drive idealisation even further than we can conceive it.* Now I come back to precision mathematics and mention a book that had a great impact as soon as it was published:

Veronese, Giuseppe: Fondamenti di geometria. Padua 1891, translated into German by *Adolf Schepp.* Leipzig 1894.

I can refer to Veronese's ideas[140] in the following way: We have always said that for us the basis for all precision geometric observations is represented by the modern concept of number and by the univocal correspondence between the points of a straight line and the real numbers. Whether one wants to accept such a foundation or not, is per se not a mathematical question, but a question of convenience. Mathematics (in a strict sense) begins only when we admit this axiom (resp. another one). *Veronese* does not do this.

He introduces the symbols $\eta_1, \eta_2, \ldots$, which stand for actual infinitely small quantities of increasing order and with which it is possible to carry out calculations

[140] See the considerations on non-Archimedean numbers in Vol. I, pp. [234]–[236] and Vol. II, pp. [220]–[224].

in a certain way. Then he thinks of an expression of the following type, where α, α_1, ... are common real numbers:

$$x = \alpha + \alpha_1\eta_1 + \alpha_2\eta_2 + \dots,$$

and it is this expression that defines, according to him, a point on the abscissa. So, for him our "rational" and "irrational" numbers taken together on the x-axis are not yet sufficient, rather he inserts points defined by actual infinitely small quantities $\eta_1, \eta_2 \dots$.

The first question that arises is: Is it possible to operate without contradictions with such expressions? This is in fact possible, so that nothing can be said against *Veronese* from the purely abstract mathematical standpoint. But we can further ask:
[157] Given that the matter is mathematically acceptable, is it also convenient to deal with it?

This brings us to the general question: *Which problems are significant at all in mathematics*?

Of course, I can theoretically pose any problem, and nobody can forbid me to deal with it. But the convenient problems are only those, which are linked to other problems about which one has to work in any case and about which one has also always worked due to the nature of things. We point to the fact that mathematics faces – within the area delimited by this formulation – sufficiently many unsolved problems, and we ask whether we should direct a great part of our clearly quite limited energies to new, at first apparently abstruse, things. Perhaps I may express myself in this manner: Provided that a deep epistemological need exists, which will be satisfied by the study of a new problem[141], then it is justified to study it; but if one does it only to do something new, then the extension is not desirable.

I wish to quote Cantor, who once said:

"*The essence of science lies in its freedom*", that is, mathematics can deal with anything one desires, as long as it draws only correct conclusions from the premises. While I theoretically accept Cantor's sentence, I add a practical restriction, which seems to me essential, namely that *everyone who has freedom also has a responsibility*. I do not want, therefore, to plead an absolute arbitrariness in the construction of mathematical ideas, but I do want to recommend to everyone that they keep in mind the whole of science.

With this, I conclude this brief digression and turn now to the field of approximation mathematics.

[141] To avoid misunderstandings, we highlight that this applies to non-Archimedean numbers (see the considerations in Vol. I and II].

III. Transition to Practical Geometry: a) Geodesy

Inaccuracy of all Practical Measurements. Examples from the Snellius Problem

In the last part we definitely spoke about precision geometry. Now, on the contrary, I pass to real *practical geometry*, that is to a type of geometry that has to actually perform geometrical operations.

We distinguish between two kinds of practical activities: Measuring and drawing (including modelling) and correspondingly between two branches of practical geometry: **[158]**

A. Geodesy (surveying),

B. Drawing geometry (descriptive geometry in the broadest sense of the word).

Our main question is: To what extent has the differentiation between approximation and precision mathematics already been accomplished?

I give the following answer:

Geodesy is the part of geometry in which the idea of approximation mathematics has found its clearest and most consistent realisation. Here one constantly studies both the accuracy of the observations and the accuracy of the results that follow from observations.

On the other hand, the *drawing geometry* is still underdeveloped from the point of view of a rational theory within approximation mathematics. In general, it is customary to point out that there are, of course, inaccuracies, and reducing them to a minimum is an essential task. The rule then is given as:

"Draw as accurately as possible, but trust the result as little as possible" (*Sebastian Finsterwalder*)[142].

First I will speak of geodesy and begin with *lower geodesy*[143], which deals with the measurement of triangles and polygons on the plane. Let us think of how the ideas of approximation mathematics come into play in the theoretical approach to measuring.

All measurements, which are here divided in two types – measurements of distances and of angles – *are subject to some inaccuracy*. This is due to different reasons, to which I will briefly refer.

[142] Translator's note: see *Sebastian Finsterwalder*, Anschauliche Einführung in den geometrischen Teil der höheren Geodäsie. Berlin 1944.

[143] Translator's note: Also called *practical geodesy*.

F. Klein, *Elementary Mathematics from a Higher Standpoint*,
DOI 10.1007/978-3-662-49439-4_9

Firstly, the limited precision of the instruments themselves has to be taken into account, because the scales used are subdivided only up to a certain limit, that is, up to a definite number of decimals of the fractions of a millimetre or of a minute. Furthermore, these scales are subject to disturbing factors (variations of the temperature, bending caused by weight forces, chemical reactions of the metal alloy, etc.). As a third factor, disturbing subjective influences that affect the observer need to be taken into account, that is, the so-called "personal equation" intervenes here. Finally, also the objects to be measured are not given exactly, because the sight lines, that are the paths of light, are not straight lines if atmospheric refraction has to be taken into account.

Due to the disturbing situations described, measurements are not precise. Furthermore, how wide the margin may be, within which the single measurement is undetermined needs to be taken into account as well. First of all, we ask ourselves *to what extent can other quantities be calculated from data that are available with a limited precision, in particular which precision will the quantities to be calculated have.*

[159] At first, one can undertake purely theoretical investigations on this issue. As a typical example we can consider the *Snellius problem of the quadrilateral*[144].

I can briefly characterize it as follows: Given three known points A, B, C (for instance three maritime signals near the coast), the problem consists in determining the position of another point P (a ship in the sea) through resection, that is by observing A, B, C from P and thus measuring the two angles $\varphi = \sphericalangle APB$ and $\psi = \sphericalangle BPC$ (Fig. 96). The theoretical problem would be: *How accurately is P determined through the two measurements, if we attribute to each angle measurement the same inaccuracy?*

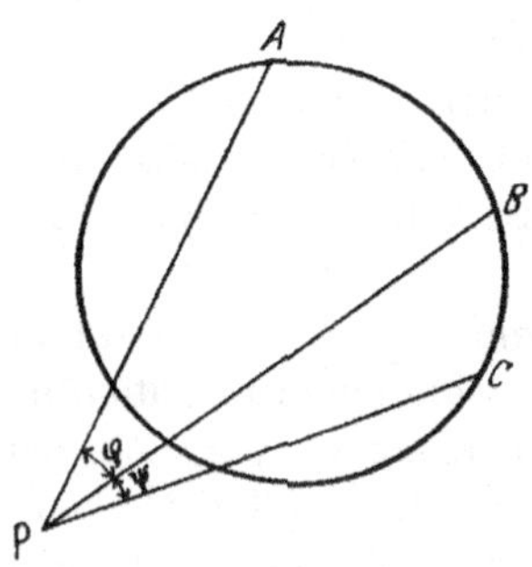

Figure 96

A circumstance here is particularly relevant; it occurs when P is located on the same circle as the three points A, B, C. In this case both the angles φ and ψ have a constant value for every position of P on the circle.

Therefore, in this case, the point P cannot be determined given the values of the two angles φ and ψ; rather it can have a totally arbitrary position on the considered circle. We express this by saying: *In the Snellius problem there exists a danger circle, namely the circle passing through A, B, C.*

144 [Often called, improperly, *Pothenot's problem*]

If you think about the fact that such a circle of indeterminacy exists, then you understand how a small inaccuracy in measuring the angles for a point P located close to this circle can be fatal. The angles φ and ψ for the points that lie close to the circle are almost the same as those obtained for a point on the circle, so that an inaccuracy in their measurement is possibly fatal for the determination of the point. So we reach the following conclusion: *In general, in proximity of the danger circle the theoretical position of a point is strongly influenced by small inaccuracies of the measurements.*

Usually, the point P is determined very imprecisely if we are near the danger circle, and with much higher precision if we are far from it.

One can even draw curves that have the same precision, which cover the whole plane, so that two points P and P', which lie on the same precision curve, are determined through the measurement of the angles φ and ψ with the same precision. These precision curves have, according to the developments by *Wilhelm Jordan*, the shapes shown in Fig. 97, from which it appears that close to the danger circle **[160]**
precision is indeed very low, but in proximity of the three targets there can be points where it is quite high[145].

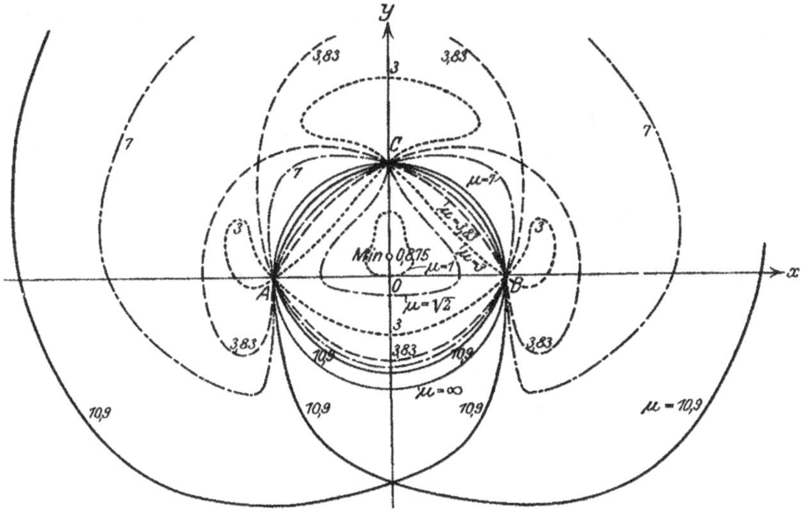

Figure 97 Precision curves in resection. ABC is an isosceles right triangle. If r is the radius of the danger circle, M the mean error[146] of the determination of the point P through resection, δ the mean error of the angle measurement, then we have $M = \mu r \delta$, where μ satisfies the equation $r^4\mu^4 = \frac{r^2+x^2+y^2}{(r^2-x^2-y^2)^2}(r^2+x^2+y^2-2ry)[(r^2+x^2+y^2)^2-4r^2x^2]$.
If μ is considered to be a constant, the equation defines a curve whose points all have the same M value, that is, a precision curve.

---

[145] See also *Wilhelm Jordan*, Handbuch der Vermessungskunde, 3rd edition, vol. 1, chapter V, Stuttgart 1888. [the above figure can be found in Jordan's work: Über die Genauigkeit einfacher geodätischer Operationen, *Zeitschrift für Mathematik und Physik*, vol. 16 (1871), pp. 397–425.]

[146] [The error of the imprecisely determined position P^* of a point P is defined as the distance P^*P.]

Determination of the Precision's Degree by Repeated Measurements. Interpretation in Principle of the Method of Least Squares

The problem just described in relation to the Snellius problem, comes back again and again in every problem of geodesy. The precision with which the angles φ and ψ have to be measured is left completely undetermined.

How can we estimate this precision? This is the next important question. I answer as follows:

In order *to reach a certain judgment about the impreciseness of the measurements*, the saving anchor can be found through *curve fitting*[147] as proposed by the *method of least squares*.

One gives as advice: Make more measurements than necessary to obtain the result and see to what extent these measurements, which in a certain sense control
[161] one another, agree with one another. By looking at how the measures fit to one another one can gain a sort of indication of probability of how precise the single measurements are and within which margins the result reasonably has to be maintained. I explain this basing the argument again on a specific example. The problem should be to determine a *quadrilateral* given its base AB (the question of the precision of the base-measurement will remain open, here). It is possible to determine the points C and D by two angle measurements both in A and in B (Fig. 98). But one adds now, moving the instrument toward C and D, the measurement of the four angles $\gamma_1, \gamma_2, \delta_1, \delta_2$, so that one obtains in total eight angles instead of only the four necessary ones. We ask, first of all, how exactly are the geometric conditions satisfied, which must exist between these angles? Obviously four conditions have to be satisfied. But, calculating with the values of the measured angles, these are accurate only within certain errors. One could say that in measuring the eight angles eight "errors" occurred; the typical procedure then is to correct the exceeding measurements so that firstly the theoretical conditions are exactly satisfied and secondly the sum of the squares of the corrections that have to be applied to the measurements becomes a minimum.

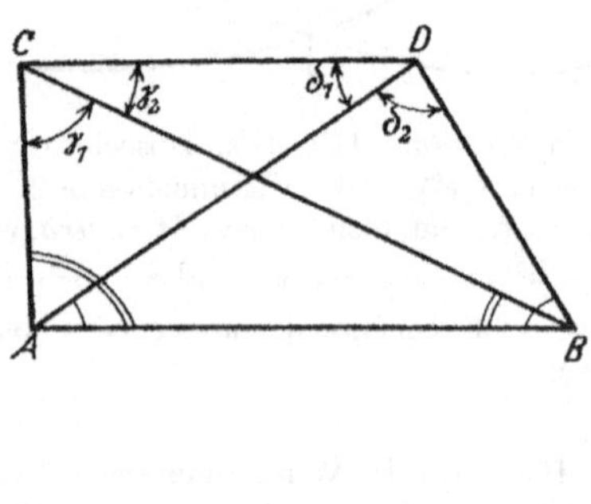

Figure 98

The justification for this method of least squares, which belongs to probability, is beyond the scope of our considerations. We soon proceed to discuss its appli-

[147] Translator's note: Literally "adjustment calculation" (Ausgleichungsrechnung).

cations: Assuming that by applying the method we have obtained two particular positions of the points C and D, are these the actual real positions of the points? This would be a wonderful method that makes it possible to obtain correct results from bad observations. However, the method of least squares must not be understood in this manner. The matter is rather the following: After having found the "most probable" position for C and D, we need to discuss the degree of precision of these positions, based on the least squares method (where of course there is a certain arbitrariness in the definition of the degree of precision; one speaks of mean error, probable error, etc.). The consequence is that around the two points C and D an elliptical region results, whose dimension is linked to the probability *that the real points C and D lie inside this region*. Thus, an additional art is needed to judge the results of the method of least squares correctly. You see, we have to deal with the systematic solution of certain tasks of approximation mathematics. The prob- **[162]** lem is not to determine the two points C and D; the question is rather which is their approximate position or within which margins it remains undetermined. *The method of least squares teaches us how to pose these questions of approximation mathematics in a definite way.*

Approximated Calculations, Explained by Means of Legendre's Theorem for Small Spherical Triangles

A further important point that I touch upon here is the question of the *kind of numerical calculation* that has to be applied in such approximation problems. If we have measured 1 m only with precision up to $\frac{1}{100}$ mm, then there is no reason to give the result with seven digits. So, my first requirement is now: *For the numerical implementation of the required calculations an abbreviated calculation is always used, that operates with as many digits as makes sense according to the concrete situation.*

For the simplification of the calculations that can thus arise, an interesting example is given by "*Legendre's theorem*" in spherical trigonometry.

It concerns the relation between a "small" spherical triangle with angles α, β, γ and a plane triangle with the same lengths of sides a, b, c and angles α', β', γ' (Fig. 99). *Legendre's theorem* says that two such triangles have angles that differ resp. by one third of the spherical excess:

$$\alpha' = \alpha - \frac{\varepsilon}{3}, \quad \beta' = \beta - \frac{\varepsilon}{3}, \quad \gamma' = \gamma - \frac{\varepsilon}{3}, \quad \text{where} \quad \varepsilon = \alpha + \beta + \gamma - 180^\circ.$$

I will not go into the proof, which is based on series expansions and the use of their lowest terms, nor into the evaluation of the error[148]. You see in any case how

[148] [The proof may be found in *Ernst Hammer:* Sphärische Trigonometrie, 4th ed., pp. 542–548. Stuttgart 1916. There you also find on pp.687–688 very interesting observations on the history of the theorem.]

useful the theorem is, because it is much easier to work with the formulas of plane trigonometry than with those of spherical trigonometry[149].

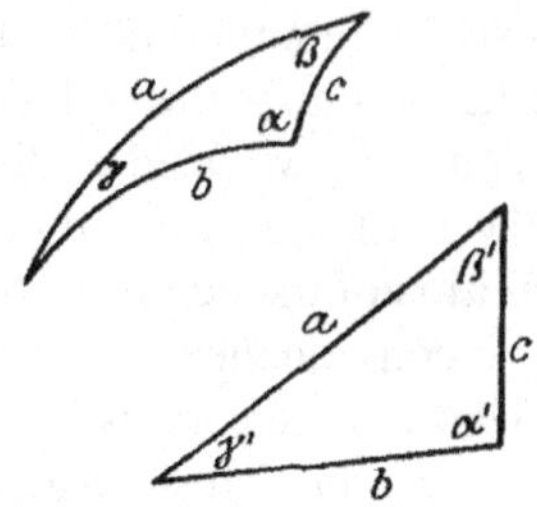

Figure 99

I summarize:

Legendre's theorem is of great help for numerical calculation and I consider this theorem a typical and elegant example of what approximation mathematics is.

Finally, some *historical remarks* deserve to be mentioned: In geodesy, the use of the method of least squares, of the abbreviated calculations, of Legendre's theorem, etc. have been used throughout starting around the beginning of the last century.
[163] *Gauß* in particular has given theoretical completion to these methods, and I will not fail to cite here a statement of *Guido Hauck* at the scientific congress (Naturforscherversammlung) of München 1899:

"Every surveyor is anointed with a drop of Gaussian oil."

This means: Every geodesist has reflected upon what approximation mathematics is and on how he reasonably subjects the empirical material to its mathematical treatment.

Pure mathematics has lost this drop of Gaussian oil since about 1860. From this period on, after the death of *Jacobi, Gauß, Poisson, Cauchy* etc., pure mathematics has turned away from these things, as if they were something lower. In this sense, the very history of the mentioned Legendre's theorem gives an excellent example. Legendre first inserted this theorem into his geometry textbook. *But in the new editions of the book it is omitted, even though spherical geometry in itself is extensively treated*[150].

So much for lower geodesy. Now I will turn to *higher geodesy*.

---

[149] [In sections 24–28 of his "Disquisitiones generales circa superficies curvas" (Göttingen 1827) = Collected works, vol. 4, pp. 251–58 *Gauß* has generalized Legendre's theorem to small geodetic triangles of any surface.]

[150] Another characteristic example that shows how the feeling for approximation mathematics got lost in pure mathematicians over the last years is given by the fact that the modern textbooks of analysis hardly touch on the *difference calculus*, which was of great importance before. Only recently has there been a slight change of direction (see Vol. I, p. [254]). Translator's note: Klein was not aware that the "new editions" of Legendre's geometry did not maintain Legendre's text, but were decisively changed and mutilated by *Alphonse Blanchet* who published revised editions from 1845 on (see *Gert Schubring*, "La diffusion internationale de la géométrie de Legendre: différentes visions des mathématiques", *Raisons – Comparaisons – Éducations. La Revue française d'éducation comparée*, 2007, 2: 31–54).

The Geodetic Meaning of Shortest Line on the Earth-Spheroid (with Postulates Concerning the Theory of Differential Equations)

Here the Earth's surface is seen as an ellipsoid and the major task is to trace *geodetics* upon the ellipsoid. The question is, to what extent the Earth's surface actually corresponds to an ellipsoid and how it is possible – despite the many irregularities of the surface (mountains, valleys, forests and meadows) – to keep on thinking about a geodetic line running along the ellipsoid.

What I have to say about this sounds very trivial, because everyone will soon consider it correct and the practitioners unconsciously already act according to it. However it is essential to clarify these aspects, in order not to introduce arguments of precision mathematics, which – as such – are not sufficient here.

What is, first of all, the precise definition of a geodetic line $x(t), y(t), z(t)$ on the ideal surface?

As it is known, we have two definitions:

One can, firstly, consider the arc length

$$\int \sqrt{\left(\frac{dx}{dt}\right)^2 + \left(\frac{dy}{dt}\right)^2 + \left(\frac{dz}{dt}\right)^2} dt$$

and require for it to become a minimum between two points to be connected. Or, we can consider the osculating plane of the curve and demand that for every point **[164]** of the curve such plane be perpendicular to the tangent plane of the surface[151]. Both definitions make use of differential and integral calculus, and these are based, as we know, on the rigorous concept of limit.

Here we have as a principal proposition:

For the precise definition of the geodetic line an exactly defined idealised surface must be given, and $\frac{dx}{dt}, \frac{dy}{dt}, \frac{dz}{dt}$ *signify limits of difference quotients.*

Having this in mind, we compare the Earth's surface to an ellipsoid.

Of course it is possible to compare at large the Earth's surface to an ellipsoid, as the works of the geodesists have shown. You can see, for instance, the reports of *Anton Börsch*[152] on the determination of vertical deflection, which speak of the comparison of central Europe, namely the best measured part of the Earth's surface, to an ellipsoidal shell. But at a "small" level the surfaces do not correspond at all. If, for instance, we want to construct the tangent plane to the Earth's surface, we need to consider – according to the ideas of precision mathematics – every variation of the surface, however small. But we do not have a smooth surface; the whole behaviour is highly irregular. If we only take into account small variations that can be observed (differences between mountains and valleys, irregularities of the ground, changing vegetation, etc.) the impossibility of the comparison becomes evident. Even more if

[151] [See, for instance, *Wilhelm Blaschke,* Differentialgeometrie Vol. I (2n ed. 1924), § 56.]

[152] Berichte über die Lotabweichungsbestimmungen. *Verhandlungen der internationalen Erdmessung*. Stuttgart 1898, 1903.

we continue – as is required by the rigorous definition – and are obliged to consider the cell structure of organisms, or the molecular structure of matter. A palpable law does not exist at all, not even a rigorous surface, but only of a space layer whose thickness looses relevance with respect to its extension. We have to say: *If we compare the Earth to an ellipsoid, this does not refer at all to "subtleties" in the behaviour of the Earth surface that can be described by difference quotients in the case of small differences, and even less by derivatives.*

And I continue, in order to express the situation even more vividly:

In practice, what the Earth's surface should be is not exactly defined at all, because it is not established which parts have to be counted in determining the mass of the Earth. In any case, if we build difference quotients of small magnitudes, these will oscillate back and forth to the utmost. In any case, you can see how we surely must not interpret the approximation as an ellipsoid. At a "smaller" level there is no
[165] correspondence. *However, this approximation works if, as I maintain, we impose a certain order of magnitude to the differences considered in difference quotients.*

Let us think of a flat terrain (a mountainous one would make the observation more difficult), where we believe to see a sharp horizon at about 10 km. We can compare the plane, passing through our eye and parallel to the horizon, with the tangent plane to the ellipsoid. More in general we can say: If, in determining the Earth's shape at a macroscopic level (which will be dealt with referring to the geodetics of the ellipsoid), we introduce $\Delta x, \Delta y, \Delta z$ and their quotients $\frac{\Delta y}{\Delta x}, \frac{\Delta z}{\Delta x}$, then we have to think of $\Delta x, \Delta y, \Delta z$, or better of $\sqrt{\Delta x^2 + \Delta y^2 + \Delta z^2}$, as pieces of a length of about 10 km. The resulting difference quotients can be seen as approximations of the derivatives of the ellipsoid, in the same sense in which the tangent plane of the ellipsoid is approximated by the horizon line that we observe. For a mountainous terrain we need, of course, to presuppose an equalizing or levelling of the mountains and valleys in order to establish similar relations. This all corresponds to what I said previously about the drawn curve, namely that if we speak of a tangent to such a curve, we do not evaluate it simply by difference quotients (or even by derivatives, which do not exist) but rather by difference quotients of a given order of magnitude.

Having agreed on this, we ask what is the usefulness of the geodetic line of the ellipsoid for practical measurements. It is clear that we cannot compare a geodetic line on the ellipsoid to the shortest line that we find empirically on the Earth's surface by considering all the identifiable details. Rather, I believe that I can suggest the following formulation, which corresponds exactly to what geodesists really do.

I want to juxtapose to the word "geodetic line" the word "*geodetic polygon*". By geodetic polygon of a given surface we can mean a polygonal chain whose sides have a given magnitude, whose vertices lie on the surface, and which is elsewhere geodetic, meaning that it is the shortest polygon of this kind between the two observed endpoints; or meaning that the osculating plane in the vertices of the polygon is always perpendicular to the tangent plane to the surface at the vertices (Fig. 100).
[166] We think at first of such a geodetic polygon on the idealised ellipsoid, and we make the sides equal to 10 km. We thus obtain a clearly determined polygon that can be

looked at within the field of pure mathematics since it was defined through definite conditions. There is, on the other hand, no difficulty in effectively producing a geodetic polygon with 10 km long sides on the empirical Earth surface – of course with the precision allowed by the circumstances. It is sufficient to sight a vertex from another one and to verify that the plane formed by two sightlines passing through the same vertex is always perpendicular to the tangent plane defined by the horizon.

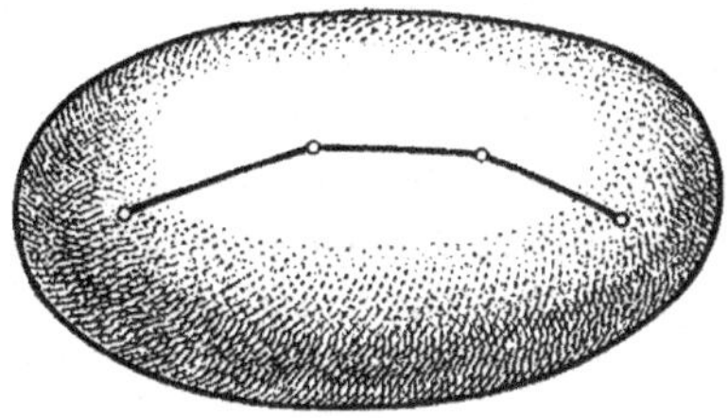

Figure 100

Now, the opinion of geodesists, which is undoubtedly correct, is at first that such an empirical polygon is almost coincident with the theoretical polygon of the ellipsoid. This decision has surely to be left to geodesists. However, they also tacitly postulate a purely mathematical theorem, which says that the polygon considered of the idealised ellipsoid can be seen as an approximation of the geodetic line of the idealised ellipsoid. This is a theorem of approximation mathematics, whose examination falls within our area. Therefore I maintain the following: To provide mathematically everything needed by geodesists it is not sufficient to define the geodetic lines on the Earth-ellipsoid with rigorous limit conditions based on the modern differential and integral calculus, but one also has to evaluate the following: How much such a geodetic line – in its course from a point A to a point B – deviates from the geodetic polygon with sides of approximately 10 km, which joins the same two points and which can be inscribed in the ideal ellipsoid. So the question would be: Is this treated somewhere in the literature or not? Is there here again a deplorable gap between the achievements of theory and the requirements of practice? *It has to be answered here that the question is in fact dealt with, but not from the points of view that are interesting to us. It should be treated at least once expressly for the present purpose.*

I somewhat generalize the question and speak of curves defined by differential equations, where I limit myself to differential equations of the first order in two variables $y' = \varphi(x, y)$ – for the sake of a simpler mode of expression. According to *Cauchy* one can deduce the existence of an integral of the differential equation by constructing a polygon with elements x, y, y' satisfying the given equations and convincing oneself that this polygon – by increasing the number of sides and decreasing the side lengths – tends to an integral curve of the differential equation. In **[167]**
this consideration of Cauchy what we are looking for is implicitly present, but put in a different light.

Cauchy makes use of the comparison between the curve and the polygons, this should mean the estimation of the differences existing between the curve and the polygon, to deduce therefrom the existence of the integral curve. We, on the contrary, will use the same estimation to obtain from it the justification (expressed directly in terms of our specific case) for approximately representing the empirical geodetic polygon on the empirical Earth surface through the geodetic line of the Earth-ellipsoid and not through the theoretical geodetic polygon! I say, in general: *What we have claimed in a specific case is still true – mutatis mutandis – whenever the question is that of treating aspects of the empirical world with differential and integral calculus, based on the exact concept of limit.* Or, in a yet different form: Considerations of the kind just described are necessary to keep pure mathematics, in its modern form that bases differential and integral calculus on the limit concept, in touch with its applications. The connection is provided by a chapter of approximation mathematics that has been quite neglected in the last decades, namely a rationally developed *finite-difference calculus.*

Now I speak, in the following appendix, of the so-called *geoid surface.*

About the Geoid and its Practical Determination

The geoid is the equipotential surface of the Earth's gravitational field; according to Newton's law of attraction, the approach is the following:

We have a potential

$$V = \int \frac{\varrho dk}{r} + \frac{\omega^2}{2}\left(x^2 + y^2\right),$$

where the integral is extended to all mass particles ρdk of the Earth's mass. The second term is present because of the angular velocity ω, with which the Earth rotates around its axis (= potential of the centrifugal force), which we identify with the z-axis. $V = \mathit{constant}$ yields the equipotential surfaces of the apparent gravity composed by the centrifugal force and the effective gravity. We obtain the force components of this apparent gravity by posing

$$X = \frac{\partial V}{\partial x}, \qquad Y = \frac{\partial V}{\partial y}, \qquad Z = \frac{\partial V}{\partial z}.$$

For the equipotential surfaces $V = \mathit{constant}$ the name "geoid surfaces" has been introduced (derived from the Greek word for Earth, which also appears in geometry, geography, etc.). Of course there exists, in correspondence to the different values of the constants, a whole series of geoid surfaces, among which one – passing through a chosen fixed point – is chosen as the main surface. The point considered for
[168] Germany lies 37 m under the height mark installed at the Berlin observatory. With an error of at most a few centimetres, this corresponds to the mean height of the sea levels of the North Sea and the Baltic Sea. The question is: How do the geoid surfaces behave, what do we know about them, what is it possible to know?

First of all, I need to report a result of recent measurements, which has changed preceding assumptions.

While presuming – in older times – a uniform distribution of the masses inside the Earth and considering then in the calculations the masses accumulated in the visible continents together with the water masses of the oceans, it was found that the geoid surfaces over the continents and the mountains strongly extended upwards. Elevations of about 200–400 m were calculated. Successive measurements of the *intensity of the gravity* have instead shown that the underlying assumption was completely wrong; it is rather the case that, generally, under the continents and the large mountains there are mass deficiencies, which balance off the excess of mass that protrudes outwards. Of course, one should not think of cavities, but instead of rock masses of a lighter kind; it is as if the visible masses of the Earth's surface reside on a liquid base, in hydrostatic equilibrium.

This is the general result, which can, however, undergo manifold particular changes. Anyway, due to these new studies, the main geoid surface appears much smoother than before. But it occasionally differs by many metres from the theoretical ellipsoidal surface, and this occurs in regions where one suspects it the least. Of course, for a large part of the Earth the important measurements are still missing[153].

Moreover, I am discussing the *means* that we have to *determine the behaviour of the geoid surfaces effectively*. I mention in particular:

a) *the astronomical determinations of the position*. These give us, for each single point, the slope of the normal, and therefore the tangent plane to the level surface or respectively to the geoid surface passing through it.

b) *the gravity measurements with the pendulum*. Gravity depends of course on the force components X, Y, Z. If we want to know the total force in the normal direction, we need to consider $P = \frac{\partial V}{\partial n}$, where n is the normal at the point considered. Performing now the gravity measurements, we measure P, so that such a measure- **[169]** ment informs how densely the level surfaces $V = \nu c (\nu = 0, 1, \ldots)$ follow one another in the neighbourhood of the observed point. Within the regions of the geoid surface that have a lower gravity the level surfaces[154] are, so to say, loosened up, wherever the gravity is greater they are huddled together (P increases and decreases in an inversely proportional way with respect to ∂n).

c) thirdly we have the *direct geodetic measurements* in horizontal and vertical directions from one point to the other. These completely determine (from a theoretical point of view) the different points of the surface at which the observation was made in their reciprocal position.

---

[153] [To learn about the developments of the matter and the related references, see, besides *Friedrich R. Helmert:* Die mathematischen und physikalischen Theorien der höheren Geodäsie, 2 voll., Leipzig 1880 and 1884, also the two articles in the *Enzyklopädie* VI 1,3: "Höhere Geodäsie" (finished in 1906) by *Paolo Pizzetti* and VI 1, 7: "Die Schwerkraft und die Massenverteilung der Erde" by *Helmert* (finished in 1910). Many references can also be found in *Richard Ambronn:* Methoden der angewandten Geophysik. Dresden and Leipzig 1926.]

[154] Translator's note: The German text says *geoid* surfaces.

So, assuming to have complete observations, one knows – together with the relative position of the single points – the tangent plane to the surface level passing through them and the derivative of the potential in the perpendicular direction. *The mathematical problem is, after having carried out all measurements with the greatest possible precision, to determine the geoid surface as precisely as possible.* In the first part of the sentence I am already requiring: If all measurements are made with the greatest possible precision; of course this is not the case, apart from single privileged countries. Therefore: The empirical material for this has been collected sufficiently only for a small portion of the Earth's surface. I will not go into the details of single observations, which are within the competence of geophysics, but I do want to stress the result of *Fridtjof Nansen*, who carried out gravity measurements on the polar ice during his expedition (1893–96) and found there a normal gravity, in accordance with what had been assumed.

Now I turn to the abstract mathematical critique in the sense of this lecture course and ask:

To what extent are the geoid surfaces ever exactly defined mathematically?

You will not be surprised by the following sentence, which I nevertheless cite expressly because among larger groups of persons the geoid surfaces are considered to be something exact, perhaps even like analytic surfaces:

Of course the definition of the geoid surface is also not a theoretically complete one, it is instead only an approximate one – as always in practical situations.

I do not want get side-tracked in subtleties, I only want to add the following:

The potential V contains at first an integral over all mass particles of the Earth. We can ask whether the air masses have to be included, etc. All this can only be established by convention; for instance, we may add to the Earth's mass the solid, liquid and gaseous masses up to 10 km distant from the Earth's surface. Thus we left out the very thin, though quite broad air masses, which extend far beyond and
[170] get lost in the medium of the outer space. In any case these should have an influence on the integral. In addition, many masses are permanently moving (circulation of water, change of air pressure, etc.), therefore the potential V becomes a function of time. Of course these temporary changes are not significant, so that we can neglect them in practice, but in the theoretical discussion we need to consider them.

I want in particular to discuss here, from my point of view, a work of *Heinrich Bruns "Die Figur der Erde"*, Berlin 1878, which is distinguished by the mathematical rigour of the exposition. I want to ask: How far does his exposition correspond to what we have always emphasized in the present lecture course?

Bruns thinks of the Earth as constituted of different layers of certain rocks, seas, etc., as schematically suggested in Fig. 101. He assumes a continuous space-filling by these masses and considers a constant density in the interior of the single layers equal to $S_1, S_2, S_3, \ldots$. We already talked about this conception when we occasionally treated the constitution of matter, and we remarked that it would be better to speak of mean values. Moreover, Bruns assumes that all the boundary surfaces of the single components are analytical, which again contradicts my opinion – according to which "analytical surfaces" do not occur at all in nature – and must therefore be understood only cum *grano salis*. The assumption that the Earth's mass is composed of homogeneous pieces of different continua, whose common boundaries are analytical surfaces, is surely not a precise description of reality, and has to be understood only approximately.

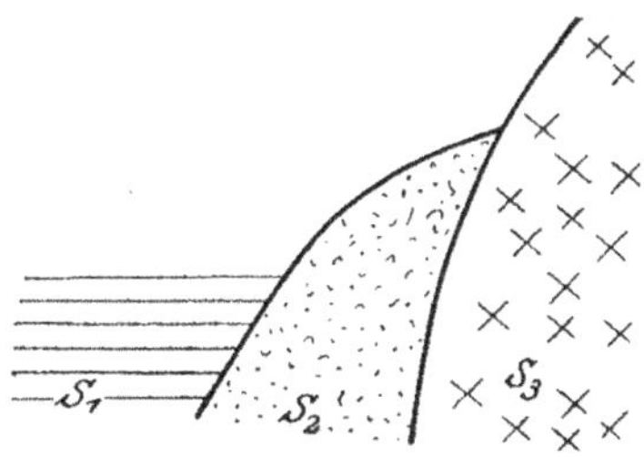

Figure 101

Now, Bruns applies the formulas and the theorems of the ideal potential theory to the assumptions mentioned and asks how do the level surfaces behave according to this precise theory of potential. He concludes that every geoid surface represents an analytical portion of surface – as long as it takes course within a same material – and when two such analytic surfaces collide at the boundary of their material layers, there is not an edge but a leap of the radius of curvature, because the ideal potential theory yields for the passage from one medium to another the continuity of the first derivative but discontinuities of the second one.

Our question is, what is the meaning of these results of Bruns or in general of all analogous theoretical developments for the practical measurements, since the assumptions – in the strict sense of the term – do not hold or, respectively, can **[171]**
hold only as idealised descriptions of the effective relations. While recalling the meaning of the geodetic lines for the measurements on the Earth's surface, I am formulating the following proposition: *We will look for an interpretation of Bruns's results similar to the interpretation obtained above with reference to the practical meaning of the geodetic lines to be drawn on the ellipsoid.*

A more detailed study would lead us too deeply into the theory of potential and must be omitted here. We conclude thus here our considerations on geodesy. Looking back we can say:

Geodesy is in general a shining example of what mathematics can do in applications and how it should be done. Of course everything is determined only approximately, but at the same time one has determined the degree of approximation in those cases in which the study can be considered as concluded.

IV. Further Considerations on Practical Geometry: b) Drawing Geometry

In the second branch of practical geometry, *drawing geometry*, things are much more awkward.

The aim of *drawing geometry* can either be the graphical representation of spatial relations (*descriptive geometry*) or the substitution of numerical calculation with drawn constructions (*graphic calculation*). In both cases our old, usual considerations are appropriate: What about precision, do theorems exist referring to it, what remains unsettled?

Postulation of a Theory of Errors for Drawing Geometry, Explained Using a Graphic Reproduction of Pascal's Theorem

I must start with the statement:

In drawing geometry, a rational theory of errors, as it exists in geodesy, has not yet been developed, with the exception of some approaches that will soon be discussed.

Here I call rational a theory of errors based on the use of probabilistic considerations, in such a way that we can evaluate the precision of a construction method by applying it again and again to the same problem, then adjusting the results obtained according to the method of least squares or otherwise.

Preliminary work towards such a theory of errors for graphical solutions can be found in the observations first carried out by the French mathematician *Émile Lemoine* and which he called geometrography. In the different solutions of a problem of construction, for instance the problem of Apollonius (to construct with ruler and compass the eight tangent circles to three given circles), Lemoine counts how many times the ruler and the compass are used. Lemoine calls the resulting number the measure of simplicity, respectively of the intricacy, of the construction. You can understand how this is linked to our observations. The larger the number is, that is **[172]**
the more often the ruler and compass are used, the more inaccurate is the drawing – generally speaking – so that we will be inclined to consider the simplest solution in Lemoine's sense as the most precise in our sense. Indeed, the number that corresponds to the "mesure de simplicité" is only a very superficial measure and the approach is therefore only a provisional one. Indeed, things that do not have the

F. Klein, *Elementary Mathematics from a Higher Standpoint*,
DOI 10.1007/978-3-662-49439-4_10

same weight are here counted as equivalent. For instance, the straight line connecting two very close points is determined in a much less accurate way than the one connecting two points which are further apart.

Whichever theory of errors of geometrical constructions one might think to have used, it is clear that it will never apply directly to "idealised" theorems of precision mathematics, but it will make use of corresponding theorems of approximation mathematics[155].

I will explain this through a characteristic example. We want to reproduce graphically *Pascal's hexagon* or respectively the related *Pascal theorem*.

It states: Given six points 1, 2, 3, 4, 5, 6 that lie on a conic section, we can trace the connecting straight lines $\dfrac{\overline{12}\,\overline{23}\,\overline{34}}{\overline{45}\,\overline{56}\,\overline{61}}$ and consider the intersections of those lines, which are one below the other in our writing; then the intersection points *a, b, c* are collinear (Fig. 102).

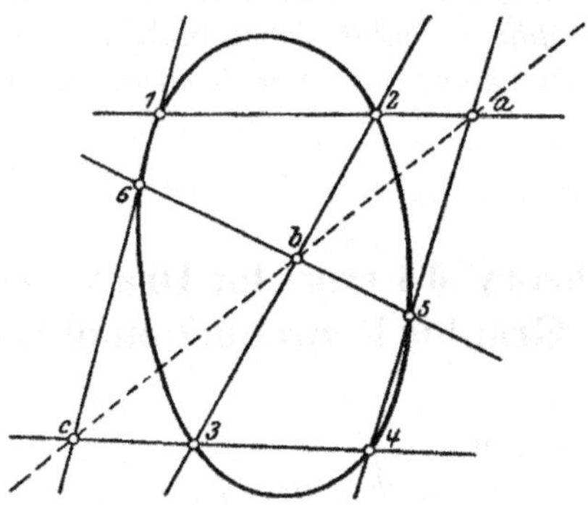

Figure 102

In contraposition to this theorem, which can be exactly proved using the usual axioms, I now ask for an approximate, extended version of Pascal's theorem[156], which at the moment I express as follows: *"If six points lie approximately on a conic section, and I draw their approximate connecting lines and trace their inter-*
[173] *section points a, b, c; then these points are approximately collinear"*. The main point consists, obviously, in specifying *everywhere the word "approximately" with precise and suitable indications of intervals.*

[155] See also *Alexander Witting*: Geometrische Konstruktionen, insbesondere in begrenzter Ebene. Programm des Gymnasiums zum heiligen Kreuz. Dresden 1899. *Witting* gives auxiliary constructions, which have the aim to avoid operations that can be performed only approximately, such as the connection of two close points through a straight line. Also see *Paul Zühlke*: Konstruktionen in begrenzter Ebene, Leipzig 1913.

[156] I briefly call this theorem "*approximate*", because it concerns questions of approximation. If the approximation is assigned with a precise margin, then the theorem becomes exact (as does every complete theorem of approximation mathematics, that is every theorem giving the interval within which the error can vary).

It is clear that it is this theorem (and not the "idealised" Pascal theorem) that acts as the real theoretical basis for the graphical construction.

The proof may be performed in a completely analytic way, which is the most convenient for the reasons explained before. First of all, we formulate Pascal's theorem analytically, that is we think of analytically proceeding through the following development:

If the six points 1, 2, 3, 4, 5, 6 on the conic have the coordinates $x_1, y_1; x_2, y_2; \ldots; x_6, y_6$, and the three resulting points a, b, c have the coordinates $x_a, y_a; \ldots; x_c, y_c$, then the following relation holds, which shows that a, b, c lie on a straight line:

$$\begin{vmatrix} x_a & y_a & 1 \\ x_b & y_b & 1 \\ x_c & y_c & 1 \end{vmatrix} = 0$$

Now we change the coordinates $x_1, y_1; \ldots; x_6, y_6$ to $x_1 + \delta x_1, y_1 + \delta y_1; \ldots; x_6 + \delta x_6, y_6 + \delta y_6$, where the $\delta x_\nu, \delta y_\nu$ are small magnitudes with respect to the other dimensions of the drawing. In order not to make things too complicated, I assume that there are no other inaccuracies, that is, the connecting lines are traced sharply and also the intersection points are marked sharply. Then it is clear that the changes of the coordinates of the intersection points a, b, c depend only on $\delta x_1, \delta y_1; \ldots$ We calculate for these changed coordinates the determinant,

$$\begin{vmatrix} x_a + \delta x_a & y_a + \delta y_a & 1 \\ x_b + \delta x_b & y_b + \delta y_b & 1 \\ x_c + \delta x_c & y_c + \delta y_c & 1 \end{vmatrix} = \Delta,$$

which geometrically represents twice the area of a triangle. *Of course, Δ no longer equals zero; it is necessary to examine if it remains small for suitable $\delta x_\nu, \delta y_\nu$.*

In geodesy and *measuring astronomy*, such a procedure is absolutely usual. These kinds of formulas are called differential formulas[157], so that we can express our requirement by saying: *We need to have at hand not only the formulas of Pascal's theorem, but also the corresponding differential formulas. Only then we will have the necessary theoretical foundation for the graphical execution of Pascal's theorem* (or better for the mathematical understanding of the graphical execution[158]). **[174]**

[157] Translator's note: *Differential formulas* are formulas which also take into account a probable error (the difference); soon Klein will explain the reason for this terminology.

[158] [A theory of precision of geometric constructions, on a different base with respect to Lemoine's geometrography, is proposed in the following works: *Ferdinand Geuer,* Die Genauigkeit geometrischer Zeichnungen (*Jahresbericht* 1902 des Progymnasiums in Durlach in Baden). *Paul E. Böhmer,* Über geometrische Approximationen (Dissertation Göttingen 1904). *Konrad Nitz,* Anwendungen der Theorie der Fehler in der Ebene auf Konstruktionen mit Zirkel und Lineal (Dissertation Königsberg 1905) and Beiträge zu einer Fehlertheorie der geometrischen Konstruktionen (*Zeitschrift für Mathematik und Physik* Vol. 53 (1906), pp.1–37). Also notable for the observation of errors is *Hans Schwerdt,* Lehrbuch der Nomographie, Berlin 1924. Finally, see the chapter "Geometrographie und Fehlertheorie" on pp. 121–129 of the work by *Theodor Vahlen,* already

[175] I want to say something more about the expression: *Differential formula*[159], so that our conscience is not troubled when using such a word.

The trouble is due to the fact that on the one hand we say that differential calculus only deals with the rigorous concept of limit, while on the other hand in the use of

---

mentioned on p. [10]. The basic idea of Geuer's work consists in the *application of Gauß's adjustment method.* Every point, resp. every straight line we are looking for, is over-determined and the adjustment is performed in a *constructive* way according to the requirement that the sum of the squares of the errors becomes a minimum. The magnitude of the corrections necessary to the adjustment gives an indication for assessing the accuracy of the drawing. The method of the *differential formulas* suggested above in the text is followed in the work of Böhmer. There we find, for instance, the differential formulas for the construction of a triangle given three sides. These formulas allow recognising the side with which the construction must start and the side with which one has to continue in order to obtain the greatest possible precision. Also for Pascal's theorem, Böhmer makes analogous considerations, aiming to find the differential formulas, but the considerations are inconclusive. These considerations are introduced by solving a problem of adjustment, namely to find the "nearest" conic section passing through six points lying approximately on a conic section. By the "nearest" conic sections he means those satisfying the adjustment conditions of Poncelet-Chebyschev, namely to minimalize the maximum distance of the given points from the conic section. Later on, also the adjustment corresponding to Gauß's minimum requirement is treated. The most successful approach seems to be that of *Konrad Nitz*, who comes back to the *theory of errors on the plane* developed by geodesists. His idea can be sketched out by the following observation. If we trace 2 straight lines represented by pencil strokes, then the intersection point is in the intersection parallelogram of the two strokes. If, in particular, the two strokes have the same thickness the parallelogram has the form of a rhombus. If we try to catch the intersection point of the two straight lines with the spike of the compass, then – assuming the validity of Gauß's error theory – the points that can be caught with the same probability are on an ellipse, whose centre coincides with the intersection of the diagonals of the parallelogram and with respect to which the directions of the sides of the parallelogram are conjugated directions. When the probability varies from 0 to 1 we obtain a bundle of similar coaxial ellipses. Among these, one, "*the mean error ellipse*" is chosen as characterizing the precision of the operation "to point with the spike of the compass at the intersection of two straight lines". If, further on, two points on the drawing sheet are realised by means of circles and if their connecting line has to be drawn with the ruler, then we find that equiprobable connecting lines constitute the envelope of a hyperbola. In the bundle of similar coaxial hyperbolas, which results if we consider all the values of probability between 0 and 1, a "*mean error hyperbola*" is determined. It serves to measure the precision of the operation "positioning of the ruler at two given points". All the constructions that can be accomplished with ruler and compass can be obtained combining five *basic constructions*, namely the two just mentioned and three more: To place the spike of the compass a) on the intersection point of a straight line and a circle, b) on the intersection point of two circles and c) to construct a circle with known centre and radius. The discussion of the errors arising from these constructions yields *three more mean error curves*. If we need now to determine the mean error for an arbitrary construction with ruler and compass, first of all we determine the mean error for each of the basic constructions used, with the help of the error curves and of the measurement of diameters of "points" and widths of stripes. Then we apply the *theory of the composition of errors*, so as to calculate the mean error of the whole construction from the mean errors of the basic constructions. In this way Nitz analysed elementary constructions, like those of the bisector of a segment, of a straight angle, of parallel lines, etc., and obtained very interesting results.]

[159] [*Alwin Walther* has delivered a talk on this topic at the Göttingen Summer Course 1926, which should be published in a largely extended version]. Translator's note: published in 1929 by Teubner, with the title "Begriff und Anwendungen des Differentials: mit besonderer Berücksichtigung der Bedürfnisse des Unterrichts und der Naturwissenschaften; nach Vorträgen beim Göttinger Mathematisch-Physikalischen Ferienkurs für Lehrer höherer Schulen Juli 1926".

differential formulas small differences are taken into account. This contradiction is solved in the following way:

Take the simple case of a function, which depends on a variable, $y = f(x)$. Then the practitioner says: To obtain $f(x + \delta x)$ one can expand the Taylor series stopping at the first term. That is, if it is

$$f(x + \delta x) = f(x) + \varphi(x) \cdot \delta x,$$

one can put $\varphi(x) = \frac{df}{dx}$. Of course, we need to handle the matter with greater care and say: $\varphi(x) \cdot \delta x$ is not the first term of the Taylor's series, but it is the remainder that we have to add to the null term; for the formula to be exact $\varphi(x)$ should not be equal to $\frac{df(x)}{dx}$ but to $\frac{df(x+\vartheta\delta x)}{dx}(0 < \vartheta < 1)$, this means that $\varphi(x)$ is not the differential quotient of $f(x)$ at x, but at an intermediate point $x + \vartheta\delta x$. *So, here a mean value for* $\frac{df(x)}{dx}$ *intervenes*. As is known, the formula

$$f(x + \delta x) = f(x) + \delta x \cdot f'(x + \vartheta\delta x)$$

is called the *mean value theorem*, so that we can say:

If we allow us to write, in some abbreviated calculation, $f(x) + \varphi(x)\delta x$ instead **[176]**
of $f(x + \delta x)$, then $\varphi(x)$ is not obtained starting from Taylor's series but from the mean value theorem.

In more complicated situations, differential formulas behave exactly as in this simple case. In practice, we express ourselves as if we were allowed to eventually omit, in Taylor's theorem, all higher powers of the increments. More precisely, the differential formula has to be seen as a remainder formula in the sense of the mean value theorem. The one does not differ much from the other, but only as long as the derivative $f'(x)$ remains almost unchanged in the interval δx.

About the Possibility to Deduce Properties of the Idealised Curve from the Empirical Shape

Now I want to pose a further, particularly interesting question. We have seen, on pp. [128] and following, that we can substitute every drawn curve with an idealised curve in such a way that not only the ordinates, but also the slope and the curvature of the empirical curve – as far as these can ever be determined from a quantitative point of view – can be approximated in a satisfactory way by the ordinates, slope and curvature of the idealised curve. *Can I now deduce from the gestalt relations of the empirical curve, which I see before my eyes, the corresponding properties of the idealised curve*?

Due to the road we follow to answer this question, my conception is necessarily decisive, that the idealised curve is something that goes beyond sensorial intuition and exists only on the base of definitions. So I cannot appeal only to intuition. Rather, we always need to reflect on whether, respectively why, things that we roughly see – so to speak – before our eyes in an empirical construction can be rigorously transferred to the idealised object *thanks to the given definitions*.

We will directly provide an example. In Fig. 103 we have a closed convex curve cut by a straight line. We observe that there are only two intersection points. Are we able to transfer this property to the corresponding idealised object?

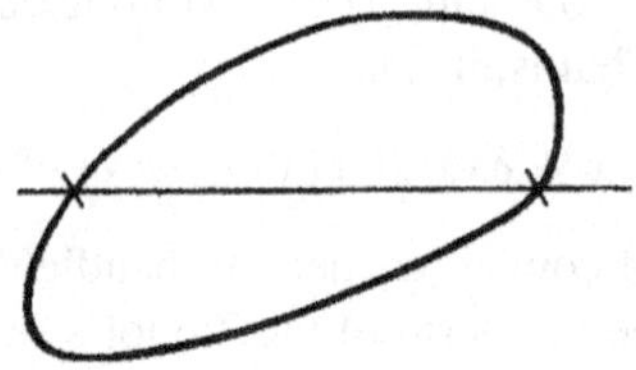

Figure 103

In our mind, we substitute the straight line of the figure by an idealised straight line, and the drawn curve – at first – by a Jordan-curve (which needs not yet have tangents, even less circles of curvature) approximating it. Intuition teaches us that the idealised straight line passes inside and outside the Jordan curve. *Therefore, according to Jordan's theorem, in the neighbourhood of the points in which the*
[177] *empirical curve and the empirical line meet there actually are intersection points of the idealised object.* However, this assertion is based on the definition by which we had earlier on introduced Jordan's curve. For the purpose of approximation we could just as well replace the empirical curve, instead of by a Jordan curve, by an everywhere dense but not closed point set (a point set which does not contain all its accumulation points, as for instance the set of rational points on the x-axis). Then the intersection points we are speaking of would not necessarily exist.

What we said gives evidence only of the *existence* of the intersection points, absolutely not of their *number*. It is possible that in the neighbourhood of the point in which the empirical figure shows only one intersection point, the idealised figure presents three or five of them (it must be an odd number, because there the idealised line passes from the exterior of Jordan's curve to the interior). This is in fact not excluded, if we do not further limit with definitions the type of Jordan curve; think of the shape of Weierstraß's curve or of Peano's curve.

To make such a limitation we decide now that the idealised curve should have at every point a defined slope and curvature, in accordance with the prototype of the empirical curve. In particular, it must be everywhere convex, as the empirical curve, and therefore it must not present an inflexion point anywhere. It is now in fact possible to *prove* that there, where the empirical figure shows one intersection point, also the idealised figure presents only one intersection point.

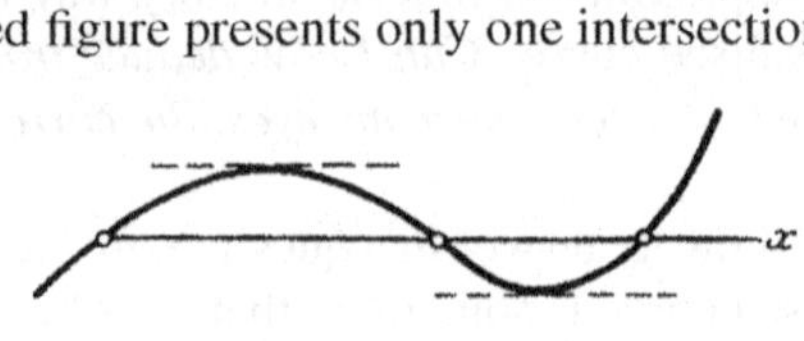

Figure 104

Indeed, if three or more intersection points were present, then we could choose the idealised straight line as x-axis (Fig. 104). The piece of curve between these intersection points would be represented by an equation $y = f(x)$, where $f(x)$ is a function whose first two derivatives exist and are continuous.

From the fact that $f(x)$ has in the observed parts of the x-axis three or more zeros it follows, thanks to Rolle's theorem, that $f''(x)$ changes its sign at least once in the interval considered. *Then our idealised curve would have an inflection point, and this contradicts the assumption made when introducing it.*

In this way, I consider the proof to be performed. *The empirical figure provides a rough orientation; it has a great heuristic value even for more detailed observations; but in the end the proof has to be based on concepts and axioms of precision geometry.*

We can find all this presented in literature in a more or less detailed form, depending on whether the author's interest lies in the rigorous foundation of certain theorems or in the discovery of new results. Both have their justification. I mention as an example of rigorous elaboration of details: **[178]**

Adolf Kneser: Einige allgemeine Sätze über die einfachsten Gestalten ebener Kurven. *Mathematische Annalen,* 41 (1893), pp. 349–376;

on the other hand, as an example of discovering many new theorems:

Christian Juel: Einleitung in die Lehre von den graphischen Kurven (Danish), *Mémoires de l'Académie des Sciences*, Copenhagen, 6. Series, volume 10 (1899), pp. 1–9[160].

Application of the Procedure on Algebraic Curves in Particular. Prerequisites Assumed in Algebra

Here I want to come back to a study that I myself conducted some years ago on the *real inflection points of algebraic plane curves*, where I made extensive use of intuitive figures[161]. The idealised regular curve that we observe is, in this case, subjected to the additional condition of being *algebraic*, excluding major complica-

[160] [We mention, in this context, the interesting lecture given by *Christian Juel at* the Naturforscherversammlung (congress of natural scientists) in Stuttgart (1906). It is printed in Vol.16 (1907), pp.196–204 of *Jahresbericht der Deutschen Mathematikervereinigung*. See also *Christian Juel*: Einleitung in die Theorie der ebenen Elementarkurven dritter und vierter Ordnung (*Mémoires de l'Académie des Sciences de Copenhagen*, 7th series, Volume XI, 1914) and the paper in Vol. 76 (1915), pp. 343- 353 of *Mathematische Annalen*; and also *Johannes Hjelmslev:* besides the treatise "Geometrie der Wirklichkeit", already mentioned on p. [15], the "Introduction a la théorie des suites monotones" (*Bulletin de l'Académie des Sciences*, Copenhagen 1914, pp. 1–74) and Darstellende Geometrie (Part 2 of the Handbook of applied Mathematics edited by *Heinrich Emil Timerding*), 1914, pp.135 and on. – Finally, the theory of ovals has to be mentioned here: *Brunn, Hermann:* "Über Ovale und Eiflächen", Dissertation München 1887 and "Exakte Grundlagen für eine Theorie der Ovale" vol. XXIV (1894), pp. 93–111, of *Sitzungsberichte der Physikalisch-Medizinischen Klasse der Königlichen Bayerischen Akademie der Wissenschaften*].

[161] [*Klein, Felix:* Eine neue Relation zwischen den Singularitäten einer algebraischen Kurve. *Mathematische Annalen*, Vol. 10 (1876), pp. 199–209, reprinted in Vol. II, pp. 78–88 (1922) of the Klein's collected mathematical works.]

tions – as will be described in the following – by requiring that we deal with curves of the n-th degree, free from "higher" order singularities.

First of all, a brief explanation about the plane algebraic curves of degree n! A plane algebraic curve C_n of the n-th degree is described by an equation of degree n in x and y:

$$Ax^n + Bx^{n-1}y + \cdots = 0.$$

Counting its addends and imagining dividing the equation by one of its non-zero coefficients, we obtain as the number of constant coefficients $\frac{n(n+3)}{2}$.

A question arises about the shape of this C_n. Of how many ovals or other com-
[179] ponents is the C_n constituted? And what about its singularities, and what about its inflection points? All these questions characterise a great chapter of the theory of algebraic curves; in particular very interesting theorems were established in this context for curves of lower degree. Obviously, here I cannot give a summary report on them, but I will confine myself to the question of the number of real inflection points.

We can write our equation briefly as $f(x, y) = 0$, or – since it is preferable for the general considerations to introduce the conceptions of projective geometry – in homogeneous form as:

$$f\,(x_1, x_2, x_3) = 0, \quad \text{where we put} \quad x = \frac{x_1}{x_3}, \quad y = \frac{x_2}{x_3}$$

If one looks for the inflection points in homogeneous coordinates one obtains the vanishing of the *Hessian determinant*

$$\Delta = \begin{vmatrix} f_{11} & f_{12} & f_{13} \\ f_{21} & f_{22} & f_{23} \\ f_{31} & f_{32} & f_{33} \end{vmatrix}$$

of the second order partial derivatives of our function $f(x_1, x_2, x_3)$. If f is of the n-th degree, then Δ is of degree $3(n-2)$, because each f_{ik} has degree $(n-2)$.

If we now consider the intersection of $f = 0$ with $\Delta = 0$, then we obtain $3n(n-2)$ intersection points – according to Bézout's theorem; *our C_n therefore has $3(n-2)n$ inflection points* ($w = 3n(n-2)$). This is the *first "Plücker formula"*[162] which leads to the determination of the singularities of the algebraic curves. More precisely, we can say:

A curve C_n has in general $3n(n-2)$ inflection points, because the curves $f = 0$ and $\Delta = 0$ have $3n(n-2)$ intersection points. But we may also investigate what happens if some of the intersection points come to coincide, in particular, how many of these intersection points are absorbed by possible singular points of the C_n.

[162] Translator's note: The *Plücker formulas* relate the number of singularities of a plane algebraic curve to its class, degree, inflection points and bitangents.

So we have, in the cases of the lowest degrees:

$$\text{for} \quad n = 2, \quad w = 0,$$
$$n = 3, \quad w = 9,$$
$$n = 4, \quad w = 24.$$

However, if we now draw a curve C_3 or C_4, we succeed in finding at most three, respectively eight, real inflection points, in fact *George Salmon* in his "Higher plane curves" hypothesised that at most $\frac{1}{3}$ of the inflection points, that is $n(n-2)$, were real. This is the question that we aspire to explore:

One should show that the number of real inflection points of a C_n cannot be **[180]**
greater than $n(n-2)$, so that in the best case only one third of the existing inflections points is real.

To the given formula for the number of inflection points I add, without proof for now, *Plücker's formula for the number of bitangents to a* C_n:

$$t = \frac{n}{2}(n-2)\left(n^2-9\right),$$

(where it is again assumed that the curve does not have so-called singular points; we will soon come back to this).

First of all we consider the *class of one* C_n, that is the number of all real and imaginary tangents to the curve through an arbitrary point not on the curve. These are all elements of the theory of algebraic curves that I need to develop here, in order to have ready the formulas necessary for the proof of our theorem concerning the number of the real inflection points. So I opt for a somewhat more detailed description.

Take the equation of the tangent at a point $x \equiv (x_1, x_2, x_3)$ of the curve, and denote by $y \equiv (y_1, y_2, y_3)$ its variable coordinates. The equation is:

$$\left(\frac{\partial f}{\partial x_1}\right) y_1 + \left(\frac{\partial f}{\partial x_2}\right) y_2 + \left(\frac{\partial f}{\partial x_3}\right) y_3 = 0,$$

where the brackets indicate that in forming the derivatives we have in mind the point x chosen as fixed. Abbreviating, the equation of the tangent at the point x in the variable coordinates y becomes:

$$f_1 y_1 + f_2 y_2 + f_3 y_3 = 0.$$

But if we fix the point y, which may not be on the curve, and consider x as the point moving along the curve, then our equation clearly defines a curve of the $(n-1)$-th degree, since the f_1, f_2, f_3 are of degree $(n-1)$ in x. This curve is called the *first polar* of the point y with respect to the curve C_n. Therefore the points x, which send their tangents through the point y, are confined on a C_{n-1}, that is we find these points as intersection points of the initial curve with the polars belonging to y. If we have determined how many such intersection points exist, then we know how many tangents there are to the C_n through y. But, in force of Bézout's theorem, this *class k of a* C_n immediately turns out to be $n(n-1)$.

Of course, this is a purely algebraic determination of the tangents, that is, we do not distinguish between real and imaginary. We still need to observe, however,
[181] that singular points of the curve may exist, which can reduce the number k because various points x can come to coincide at such a singular point and we do not want to count the connecting line xy as a tangent in this case.

So the question is: How are the formulas simplified, if *singular points* occur? What are singular points in the first place? Here as well, I describe only the most elementary prerequisites, strictly what we need. So we define straightforwardly:

We have a singular point if for a point x the first derivatives f_1, f_2, f_3 *vanish at the same time, that is, the equation of the tangent at this point is satisfied by any point y.* We have a higher singular point if the second derivatives or, with these, also the third ones etc. are either zero or satisfy particular algebraic conditions, such as $f_{11}f_{22} - f_{12}^2 = 0$.

The case of the singular points can be the easiest understood by placing the vertex $x_1 = 0, x_2 = 0, x_3 = 1$ of the coordinate triangle in the singular point (Fig. 105). The equation of the curve is then simplified as follows:

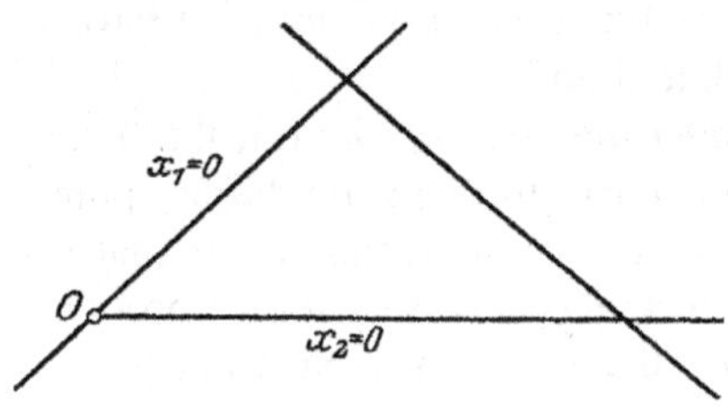

Figure 105

First of all, each equation of degree n can be ordered according to the powers of x_3 as follows:

$$f(x_1, x_2, x_3) = x_3^n \cdot \varphi_0 + x_3^{n-1}\varphi_1 + \cdots + x_3^0\varphi_n = 0,$$

where $\varphi_0, \varphi_1, \ldots, \varphi_n$ are homogeneous functions of degree $0, 1, \ldots, n$ in x_1, x_2. Now, if the point $x_1 = 0, x_2 = 0$ has to belong to the curve, the constant φ_0 (the term of dimension zero) must vanish. If the point is singular in the sense of the definition just given, then also the linear terms must vanish, so that the series expansion begins with the quadratic terms:

$$0 = x_3^{n-2}\varphi_2 + \cdots + x_3^0\varphi_n.$$

If we write down the explicit value of φ_2, one might obtain:

$$\varphi_2 = a_{11}x_1^2 + 2a_{12}x_1x_2 + a_{22}x_2^2.$$

As a particular case it can occur here that φ_2 vanishes identically, and as another case that $\varphi_2 = 0$ leads to a double solution for $\frac{x_1}{x_2}$. But if we do not further specify the coefficients of the curve other than as we did in the premise, namely choosing them so that a singular point exists at all, then the equation $\varphi_2 = 0$ yields two different values for the ratio $\frac{x_1}{x_2}$. The consequence is that our curve of the n-th degree has a "*double point*" in zero, through which two "*separate branches*" pass (Fig. 106). **[182]**
We want to limit the entire discussion we intend to carry out to the lowest kind of singular point (the "*ordinary*" double point).

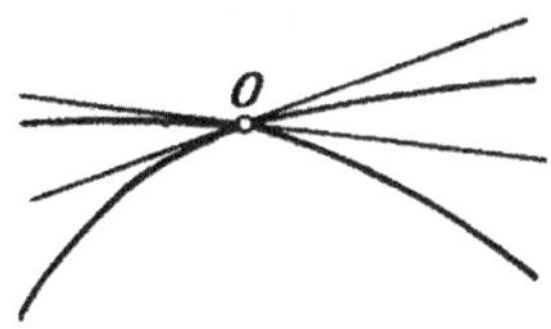

Figure 106

It is easy to see what the effect of such a double point is on the class k of the curve:

The equation of the first polar of the point $y \equiv (y_1, y_2, y_3)$ is

$$f_1 y_1 + f_2 y_2 + f_3 y_3 = 0.$$

Now we put $f = x_3^{n-2}\varphi_2 + \cdots + x_3^0\varphi_n$. We immediately see that the first polar passes once through the double point $x_1 = 0, x_2 = 0$, with one branch whose slope differs in general from the slope of both tangents to the double point, given by $\varphi_2 = 0$. Therefore, the double point absorbs also two of the intersection points of the initial curve with the first polar, so that we have the theorem:

If the initial curve has d ordinary double points, the class k of the curve is reduced by 2d, because at each of the double points two intersection points of the initial curve and the polar coincide, and we do not want to count the line joining the point y (taken arbitrarily outside the curve) with a double point x as a tangent of $f = 0$.

This is all I need from Plücker's formulas, beyond the initial statements. So I omit the further questions *Plücker* has studied: How many inflection points and how many bitangents are absorbed by a double point? Rather, let me add the following observation:

Suppose f is actually written down, with arbitrary coefficients, the same is done for the equations $f_1 = 0, f_2 = 0, f_3 = 0$ (from which it follows, thanks to a well known theorem by Euler, that $f = 0$). Will the latter be generally compatible, or which conditions need to be posed on the coefficients of f in order to obtain this, that is, in order to get a singular point?

In the case of a C_2 we can still write this down easily. We have

$$f = a_{11}x_1^2 + 2a_{12}x_1x_2 + a_{22}x_2^2 + 2a_{13}x_1x_3 + 2a_{23}x_2x_3 + a_{33}x_3^2.$$

It follows, since $f_1 = 0, f_2 = 0, f_3 = 0$

$$\frac{1}{2} f_1 = a_{11}x_1 + a_{12}x_2 + a_{13}x_3 = 0,$$
$$\frac{1}{2} f_2 = a_{21}x_1 + a_{22}x_2 + a_{23}x_3 = 0,$$
$$\frac{1}{2} f_3 = a_{31}x_1 + a_{32}x_2 + a_{33}x_3 = 0.$$

These are three homogeneous equations in x_1, x_2, x_3, that are compatible only if the condition

$$\Delta = \begin{vmatrix} a_{11} & a_{12} & a_{13} \\ a_{21} & a_{22} & a_{23} \\ a_{31} & a_{32} & a_{33} \end{vmatrix} = 0$$

[183] is satisfied. So we have the well-known theorem:

If a conic section has a singular point then the determinant of the coefficients has to vanish.

For the curves of higher degree the case is exactly analogous. Here $f_1 = 0$, $f_2 = 0, f_3 = 0$ yield three equations of the $(n-1)$-th degree, which should be mutually compatible. To this purpose the resultant or, as is usually said, the *discriminant D of the initial curve* must vanish. In the particular case of the curve C_2 the discriminant becomes equal to the determinant Δ. Underlining the word used above "in general", I can say:

In order for a C_n to have a singular point, the discriminant of its coefficients must vanish; so that in general the curve is non-singular (i.e., insofar as the coefficients do not satisfy the condition mentioned).

We can state an analogous theorem for higher order singularities; I am connecting forthwith a further conclusion to this, by saying: *In order for the C_n to have even higher order singularities, the coefficients of f must fulfil, besides $D = 0$, other algebraic conditions; therefore a curve that has $D = 0$, has at the point considered (x_1, x_2, x_3) in general only an ordinary double point, which reduces the class of the curve by 2 units; only additional conditions can ensure the existence of higher order singularities.*

This is the general algebraic basis, within which we do not yet distinguish the real from the imaginary, and in which we can even consider complex coefficients for $f(x_1, x_2, x_3)$.

Establishing the Theorem We Want to Prove: $w' + 2t'' = n(n-2)$

Now I go on to analyse *which reality theorems can be established based on the general algebraic theory*. First of all, we need to introduce a series of conventions. We suppose $f(x_1, x_2, x_3)$ has real coefficients or rather real variables, so that we can immediately tackle the totality of curves C_n, and ask what can be said generally about the gestalt of the curves C_n.

Furthermore, we distinguish the number of inflection points into two classes, real and imaginary; since the latter always occur in pairs, we put $w = w' + 2\overline{w}$.

Similarly, we divide the bitangents into two classes, real and imaginary. But here we divide the real bitangents again into two categories, thinking about the following distinction: The two contact points of a bitangent can be real; but they can also be imaginary, in which case we see an isolated bitangent. So: *The real bitangents are divided into those with real contact points, whose number we call t', and those with imaginary contact points (isolated bitangents), whose number we call t''. So we* **[184]** *have as the total number of bitangents t:*

$$t = t' + t'' + 2\overline{t},$$

where $2\overline{t}$ is the number of imaginary bitangents.

We have something similar in the case of double points. Here we will write:

$$d = d' + d'' + 2\overline{d},$$

where $2\overline{d}$ is the number of imaginary double points, which are not considered when undertaking the studies of reality, d' is the number of real double points with real branches, and d'' the number of real isolated double points.

These all are only general conventions on how we will use certain letters. We now ask, firstly, for the *total number of shapes of the curves C_n*, whether they have double points or not.

I recall, first of all, the *shapes for the case $n = 2$*.

If there is no double point, we have either a real or an imaginary curve. This kind of expression means that we see the three types of real curves, ellipse, parabola and hyperbola, as not significantly different, provided that each type can be obtained from the other by means of a projective transformation. But if we admit double points, then the C_2 will degenerate into a real or imaginary pair of lines; in the last case our eye will see one isolated point. In these cases the forms can be generated in a simple way from the hyperbola, respectively, from the ellipse by a limit passage.

The ellipse, parabola and hyperbola are called curves of *even* character, because a straight line cuts them either at two points or at none.

Now, about the C_3. Here we always have three or one intersection points with a straight line, and it necessarily appears to be a so-called *odd* curve, like the one drawn in Fig. 107 (with its asymptotes). Next to that, in addition, there can be an even branch.

This distinction can be extended to curves C_n, for which we obtain the following, somewhat uncertain representation:

A curve C_n without singular points is composed of a finite number of closed branches. If n is even, then we have only even branches, if n is odd, next to possible even branches, there is one odd branch (two or more odd components cannot occur, because they would mutually intersect, so that the curve would not be free from singularities, contrarily to what was assumed).

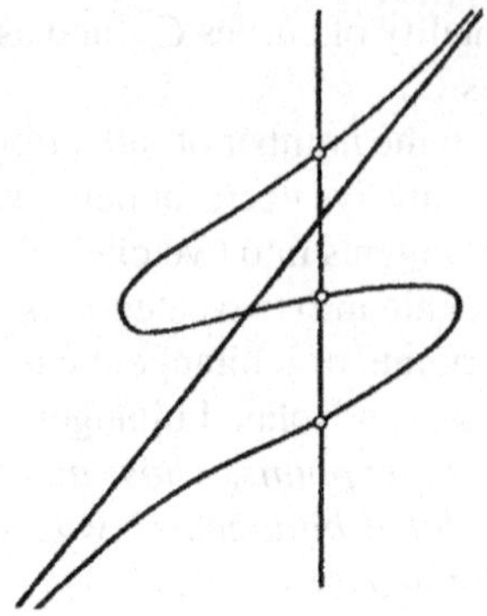

Figure 107

[185] What this looks like in individual cases has to be left to a specific description, or to deeper investigations. The beautiful thing is, however, that the theorem on inflection points, which I will present now, does not presuppose particular specific knowledge. Indeed, I assert:

Independently from the various forms that a non-singular curve C_n may assume, and that may actually be very multiform as n increases, the following relation holds for the real inflection points and the isolated bitangents:

$$w' + 2t'' = n(n-2).$$

Now we want to deal in detail with the proof of this theorem.

It is clear that, if the theorem holds, $w' \leqq n(n-2)$ is true, and that therefore the result concerning the real inflection points found by Salmon by induction is correct, namely: *From the total number of inflection points, which a curve C_n has according to Plücker, at most one third can be real.*

Principles of the Continuity Proof to be Presented

To prove our theorem we preliminarily develop for ourselves an auxiliary intuition.

The equation of a C_n has, as we have seen, $\frac{n(n+3)}{2}$ constants. We consider these as coordinates of a point in a higher dimensional space and call this point the "*representational point*"[163] belonging to the curves C_n. If the C_n assumes all possible shapes, that is if the coefficients vary arbitrarily in $f = 0$, the representational point varies in the *whole* $\frac{n(n+3)}{2}$*-dimensional space*. You will see how useful it is to choose for support such a space and to consider it alongside all the C_n. In this way some reflections about manifolds are simplified. Indeed, we are used to making considerations on continuity exclusively in point spaces.

[163] Translator's note: Such a point is said to vary in the *dual space* of the linear space of the C_n, that Klein calls *representational space*.

Now, amongst all the curves C_n, we study those curves, for which the discriminant D vanishes, that is, for which there is at least one singular point (possibly more). Being $D = 0$ an algebraic equation in the coefficients of $f = 0$ (it was obtained by elimination from other algebraic equations), we cut – thanks to $D = 0$ – from the representational $\frac{n(n+3)}{2}$-dimensional space an algebraic manifold of dimension $\frac{n(n+3)}{2} - 1$, that I call *surface*, transferring here the usual terminology from R_3 to $R_{\frac{n(n+3)}{2}}$.

What is the shape of such a surface in the higher dimensional space? Since there are already so varied shapes among the curves on the plane, major complications might occur here. However here things are easy, because we are dealing with alge- **[186]** braic structures. I immediately give the following theorem:

An algebraic surface pervades the space with a finite number of walls and divides it into a finite number of chambers; the chambers meet at walls, which have one dimension less than that of the given space. This holds, in our representational space, in particular for the surface $D = 0$.

What kind of plane curves of the n-th degree are those for which the discriminant D equals zero?

We consider at first curves with an ordinary double point, which reduces the class by two units, and then curves with higher order singularities (among which I also count the simultaneous occurrence of several double points). But if such higher order singularities should occur, besides $D = 0$ also other algebraic conditions need to be fulfilled, but these are available only in a finite number. This results from the fact that, as long as n itself is finite, these conditions will concern a finite number of distinctions: whether some zeros are coincident or undetermined, etc. It is for our following observation particularly important that here only a *finite* number of such conditions is possible.

Which are the limitations for the representational point due to these conditions for the higher order singularities? I immediately give the following theorem: *The representational points of those algebraic curves that present higher order singularities determine on the surface $D = 0$ at most a finite number of "algebraic curves", that is algebraic manifolds whose dimension is lower than $\frac{n(n+3)}{2}$ by at least two.*

This can be represented more vividly using a spatial description: The algebraic curves with higher order singularities trace on the walls $D = 0$ that separate the different chambers of the higher dimensional space, so to say, ornaments or decorations, which nevertheless do not fill up the walls, and which can rise at most to the dimension $\frac{n(n+3)}{2} - 2$ (see the schematic Fig. 108).

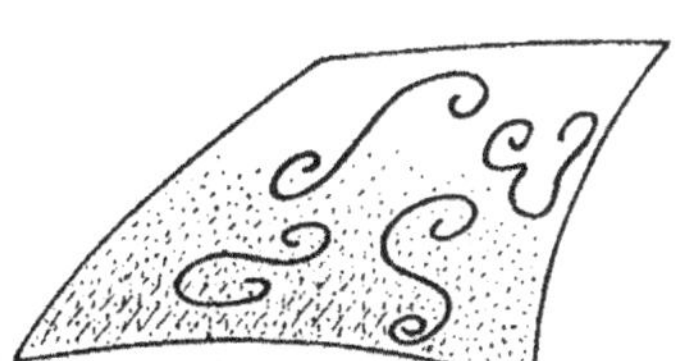

Figure 108

You can use the following concrete picture. Take R_3 and consider as walls the six faces of a cube. For these we should have $D = 0$, but the edges of the cube would correspond to the higher order singularities. This is the idea to which the following theorem, fundamental for the further considerations that will follow, is connected.

[187] We want to deal with a non-singular curve, thus considering a point outside the surface $D = 0$ in the representational space. By means of continuous changes of the coefficients of the algebraic curve, from this point we reach another point; if we want to reach a given endpoint we might need to cross a finite number of walls but we do not have to necessarily cross the ornaments on them (Fig. 109). Expressed differently:

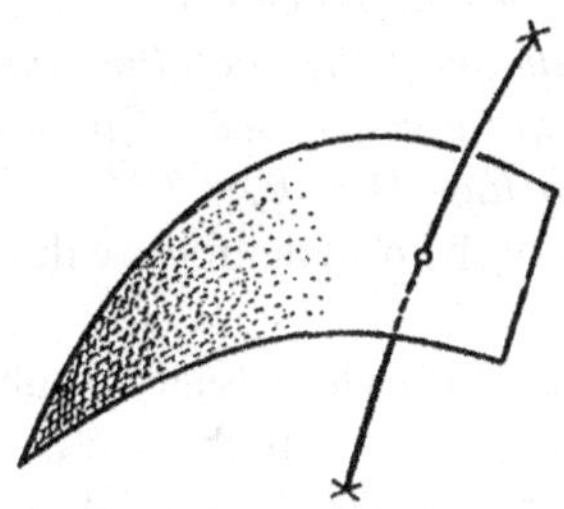

Figure 109

It is possible to pass continuously from any non-singular curve to any other such curve by interposition of other curves in such a way that during this procedure one will meet no other occurrence of singularities, apart from a finite number of times a curve with an ordinary double point, no matter whether the curve has at that point real or imaginary branches.

The theorem established thus, or respectively the point of view on which it is based, is the core of our consideration. We pose the question: is this theorem compatible with our previous discussions about the concept of curve?

If we had not had to do with algebraic surfaces but with arbitrary surfaces, then the conclusions would not be admissible, as already shown by the example of the Peano curve, which completely fills a part of the plane. But if we impose further conditions upon the concept of curve, then we already have with the Jordan curve, an example of such a curve that divides the plane into an inner and an outer region. But here we have to do with algebraic curves and surfaces, so that – even if a deeper reflection is surely desirable – we can simply claim that:

The general relations of connectedness of multi-dimensional spaces, of which we are speaking, namely that a surface acts as a wall, but not a curve, will surely be valid for algebraic curves and surfaces.

But there still remains the question of why it was possible before, when we spoke of the number of chambers, to use the word "finite". Admitting that the walls divide the space into chambers, why should the number of these be finite? This is a further point whose need could be questioned. I say here as follows: If there were infinitely many chambers, then there would be an *accumulation point* for the singularities, and this is incompatible with the properties of the algebraic functions. I cannot explain this in detail, but I repeat again what was already said:

Our first consideration concerns whether the space is at all divided into separate **[188]**
chambers. The second consideration concerns the fact that the number of these is finite.

Transition of a C_n to a Shape with a Double Point

Thus, having the most general geometric background for the further developments, we can ask:

How does the transition – by continuous variation of the coefficients – evolve, in the real field from an algebraic curve without double points to a curve with an ordinary double point (which reduces the class only by two units), and then back to a non-singular curve?

First of all, I draw a figure merely empirically, using it as a guideline and then trying to obtain from it exact theorems about the algebraic curve. In this way we apply the characteristic shift in the thought process shown on p. [176].

In Fig. 110 we have the following situation: The curve with a double point arises when the two branches, free from double points, of curve 1 approach each other from above and below, and merge into a double point, thereafter at curve 2 separating again to the right and left. All that is characteristic in this figure will remain so in the idealised algebraic curve. Our question concerns only those things, which can no longer be grasped intuitively with the limit passage, because they become – so to speak – microscopic.

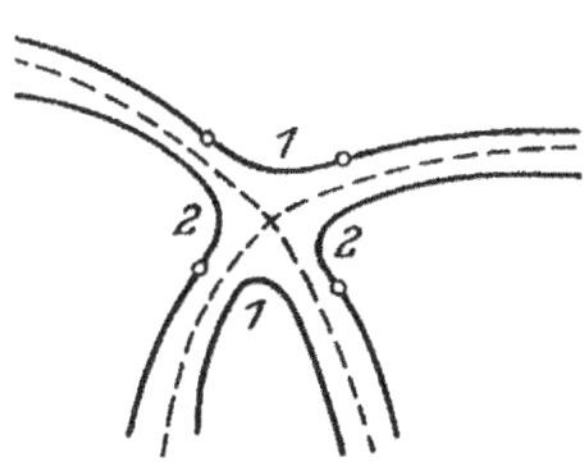

Figure 110

However with the figure above I have illustrated only one of two possible situations. A second possibility, that has the same right, is that an oval shaped curve contracts to an isolated double point, then disappearing, as explained by Fig. 111.

Now we ask, still remaining in the empirical field, for the inflection points, or – as we will also say – for the inflections. From the first figure we can read off the following remarkable fact: *At the moment when it arises the double point absorbs two real inflections and releases them when it vanishes.* Or expressed a bit differently: The curves 1 and 2 both have two inflections near the double point, which at the limit will collapse into the double point itself. On the other hand, it is clear from the second figure that:

If an oval contracts into an isolated double point and then vanishes, real inflections are not involved at all, if the situation is as shown in the figure.

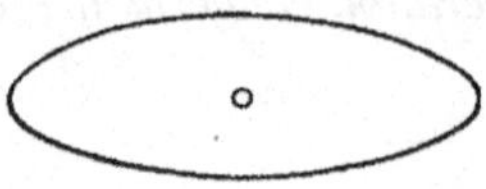

Figure 111

[189] What can we say now about the corresponding aspects of the algebraic idealised curves? In particular, to link back to the last statement: If in an algebraic curve an isolated double point arises by continuous changes of an oval, does in general the oval have no real inflections immediately before, and why does it not? Here "in general" means that we exclude the presence of higher algebraic conditions for the coefficients of $f = 0$, except for $D = 0$. So, in particular, the double point should reduce the class of the curve only by two units. We must now investigate whether in the pertinent case a *snake-like oval*[164] of the algebraic curve possibly contracts into a double point (Fig. 112).

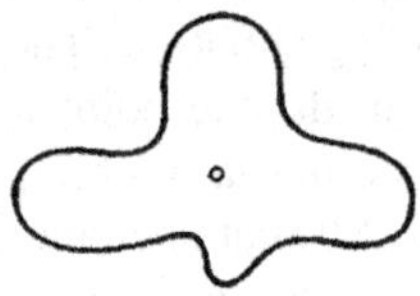

Figure 112

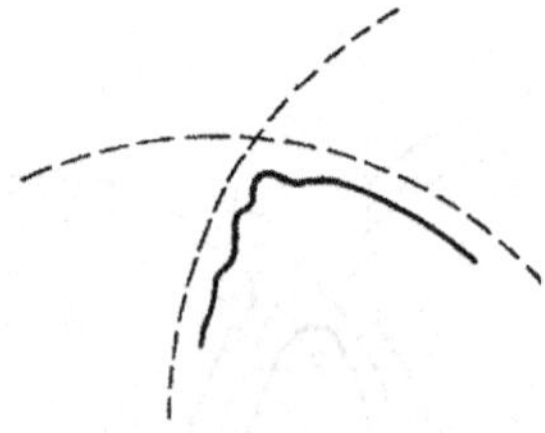

Figure 113

Similarly, with respect to our first case we can say: Also in the case of a double point with real branches, besides the two inflections, which surely exist as our figure shows, possibly still more inflections might move into the double point; or arise from it (Fig. 113). The question is whether one can claim that this is generally not the case. *I give the necessary proof in the case of the oval, because it is the most convenient for this purpose.* I take an oval with two inflections, and claim referring to it:

[164] Here and in the following the term "oval" stands for an even branch of a curve, not only for an everywhere *convex* closed curve.

If the oval has two inflections, then there is a point y outside the oval from which we can construct four tangents to the oval (Fig. 114).

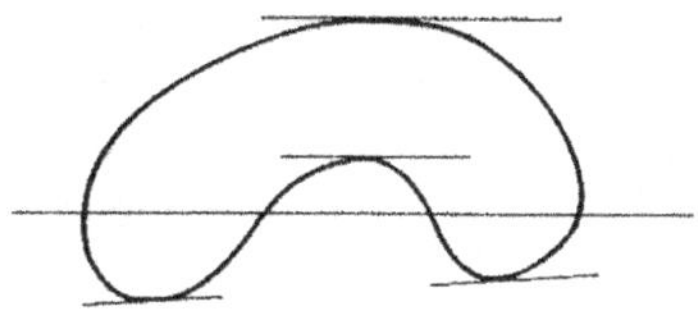

Figure 114

First of all, we can find a straight line that cuts the oval at four points. It is sufficient to connect with a straight line two points of the convex branch that lies between the two inflection points. If we take this line as x-axis and observe the ordinates that are perpendicular to it, it is evident that each of the four parts in which the oval is subdivided by the x-axis has at least one maximum or minimum of the ordinate, and thus our oval has four tangents parallel to the x-axis. The point at infinity of the x-axis is therefore a point through which at least four real tangents to the oval pass. With a slight modification of this construction we can obtain an **[190]** entire domain of such points p.

And now we arrive at a contradiction. If, namely, our oval contracts into an isolated double point (maintaining its inflections until the limit) then for all points p considered the four tangents clearly come to coincide with the connecting line $\overline{xp}$; therefore, for these points p, our double point absorbs four real tangents. We deduce that it absorbs four tangents for *each* point p, speaking algebraically, while we expressly assumed that our double point would reduce the class only by two units. Consequently, we cannot accept an oval that has two or more inflection points before the limit passage.

We can do something completely analogous for a double point with real branches (Fig. 115):

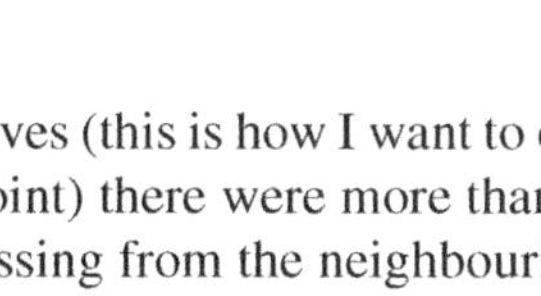

Figure 115

If in the neighbouring curves (this is how I want to call the curves in proximity of the curve with the double point) there were more than the two necessary inflection points, then here too – in passing from the neighbouring curves to the curve with a double point – the class would be reduced by more than two units, contrary to our explicit assumption that the reduction be of only two units. So we have proved the following theorem:

If an algebraic curve gains, by continuous changes of its shape, a real double point that reduces the class by two units, one needs to distinguish the cases in which the branches through the double point are real or imaginary. In the first case the double point absorbs two real inflections, in the second case no real inflection.

The deduction of this theorem is, according to the meaning of this lecture course, in some way central to our reflections about the inflection points of the algebraic curves because there the comparison between what appears intuitively and what follows conceptually from the definitions is most clearly established.

Examples of Curves for Which the Theorem Holds; Case of Even *n*

Now I turn to a simpler part of our development, in which our task should be *to construct examples of algebraic curves in which the relationship* $w' + 2t'' = n(n-2)$, *that we want to prove, is effectively satisfied.* Once we have examples, we can convince ourselves that the relation is generally valid by verifying that it is still valid after the most general continuous change that we can observe.

The construction of the examples we are talking of is easiest if we begin with
[191] the construction of curves with as many double points as possible, which are decomposed into lower-degree components, so that one can have an overlook on the whole shape of the curve; moreover, one can pass from the curves thus obtained to non-singular neighbouring curves, which will then satisfy our relationship.

It is convenient to distinguish here between even and odd n. *Let n be even, at first.* We start by choosing $n = 4$, where it is still easy to draw what I want to show you.

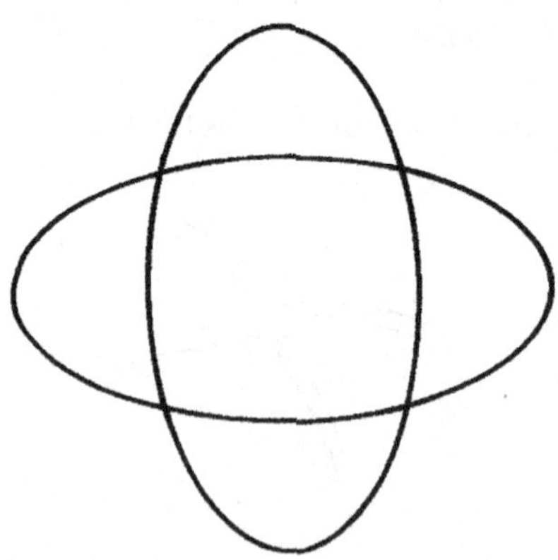

Figure 116

We simply draw two congruent ellipses Ω_1 and Ω_2 with the same centre, rotated one with respect to the other by 90°, which meet in four real points (Fig. 116). Then $\Omega_1\Omega_2 = 0$ is our degenerate[165] C_4. It clearly has four real double points, namely

[165] Translator's note: In the subject index a different term is used, namely *decomposable* (or *reducible*) *curve*.

the intersection points of the two ellipses; moreover, algebraically, the curve cannot have more than these four double points (according to Bézout's theorem). Therefore all the double points of the curve are real and with real branches.

The class of our curve is $k = 4$, since every ellipse is of class 2. For the general C_4 there are $4 \cdot 3 = 12$ tangents to the curve from an external point y. Therefore, there has been a reduction of the class by $12 - 4 = 8$ units; it follows that each of the four double points of our C_4 has reduced the class of the C_4 by two units.

We now study the bitangents of our C_4. According to the general theory, it must be $\frac{n}{2}(n-2)(n^2-9)$ bitangents, that is 28. These 28 bitangents can be represented in the figure by the four tangents common to the two ellipses, but also by the six straight lines connecting the four double points, each of which has to be counted four times. This last fact can be deduced first by introducing the six connecting lines with multiplicity α, so that they represent 6α bitangents, and then by considering that, besides these connecting lines, there are yet four more tangents common to the ellipses, which are bitangents, and then by putting $4 + 6\alpha = 28$. By the way, we will soon find further confirmation for the multiplicity 4. Adding, moreover, that our particular C_4 has no inflection point, we know everything about it for our purposes. Now we move to a *neighbouring* curve, in that we put $\Omega_1\Omega_2 = \pm\varepsilon$, where ε is a small positive quantity.

To do this, we can write something like: **[192]**

$$\Omega_1 = \frac{x^2}{a^2} + \frac{y^2}{b^2} - 1, \quad \Omega_2 = \frac{x^2}{b^2} + \frac{y^2}{a^2} - 1.$$

And we see that: Ω_1 is smaller than zero in the interior of the ellipse Ω_1, greater than zero in its exterior, the corresponding holds for Ω_2, so that $\Omega_1\Omega_2$ becomes negative in the regions that are not common to the two interiors of the ellipses (see Fig. 116).

If we now first choose the negative sign $\Omega_1\Omega_2 = -\varepsilon$, we need to draw a non-singular neighbouring curve that runs entirely in those regions. This happens if we place four ovals in them (Fig. 117).

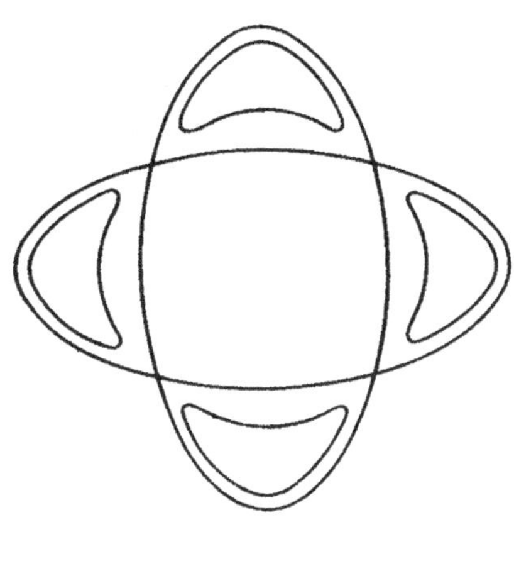

Figure 117

You can see that we obtain for our neighbouring curve exactly eight real inflections ($w' = 8$), since it must be that every double point produces two inflections, according to our general theorem, and that the ellipses have, by their nature, no inflections.

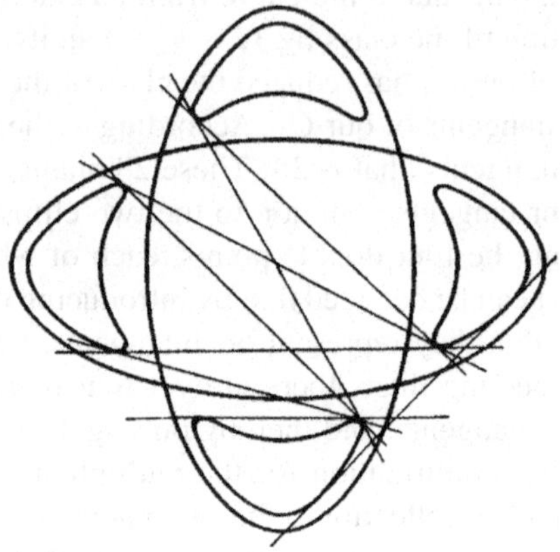

Figure 118

However, what is the situation like for the 28 bitangents of the curve thus constructed? In this case all of them can be verified easily (Fig. 118). First of all, the four tangents mentioned, common to both ellipses, evidently yield four tangents common to our ovals, that is four real bitangents; moreover, each of the six lines connecting two intersection points of the ellipses splits into four real bitangents, as shown in the figure, so that we obtain, indeed, the 28 existing bitangents. *Our curve does not have isolated bitangents*, because – besides the 28 just constructed, which all have real contact points – the general theory does not admit any more bitangents. Thus $t'' = 0$ and we obtain indeed:

$$w' + 2t'' = 8 = 4 \cdot 2 = n(n-2).$$

This means that our first example satisfies the relationship.

[193] The situation is somewhat different if we choose the second neighbouring curve as $\Omega_1 \Omega_2 = +\varepsilon$; the result is nevertheless the same $w'' + 2t''$, as far as the sum is concerned.

The curve is composed of two branches, an exterior one with eight inflections and an interior one without inflections (Fig. 119). This kind of curve, in which one branch completely encloses the other one, is called a *belt curve*.

First of all, here, too, each double point has produced two real inflections. Moreover, here, too, the four tangents common to the two ellipses have become actual bitangents. As for what concerns the remaining 24 bitangents, which the curve should have according to the general theory, they are necessarily imaginary. If, for instance, we draw the line $\overline{24}$ connecting the intersection points 2 and 4 of our original ellipses, from which four bitangents should arise, then we see that, even if we move it slightly toward its original position, $\overline{24}$ always has with the curve four totally distant real intersection points, which are not pairwise coincident. So, if there are bitangents "in the neighbourhood" of $\overline{24}$, these are necessarily imaginary. So we can conclude:

Our belt curve has four real bitangents each with two real contact points and, moreover, 24 imaginary bitangents, but not isolated bitangents.

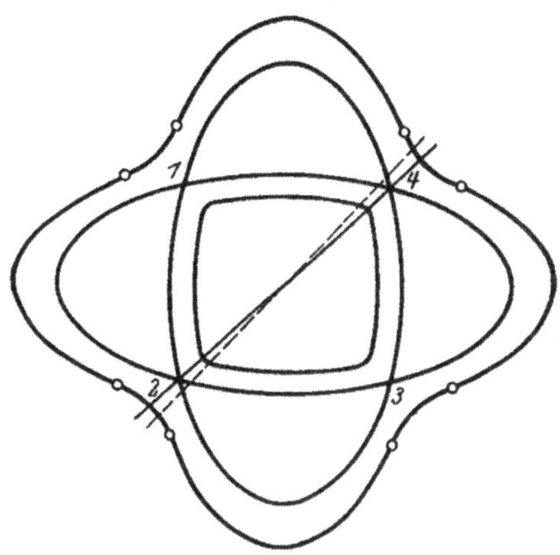

Figure 119

In our notation it is therefore $t'' = 0$, and again the relationship $w' + 2t'' = 8$ is fulfilled.

With these considerations the case $n = 4$ is completed, for the present, that is in the case $n = 4$ we have constructed examples of curves for which our relationship holds. To obtain the same for even higher values of n, first of all we construct $\frac{n}{2}$ congruent ellipses with the same centre, so that consecutive major axes are mutually inclined by an angle of $\frac{2\pi}{n}$. If we take, for instance, four ellipses as in Fig. 120, we obtain a curve C_8. In the curve C_n that we have constructed this way we can count in the same way as we did before for the C_4.

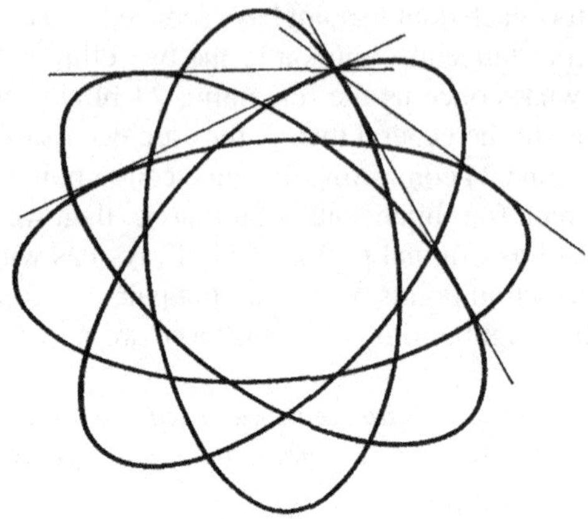

Figure 120

[194] 1. *How many double points* are there?

Each ellipse intersects each of the others in four real points, so we have in total

$$\frac{4\frac{n}{2}\left(\frac{n}{2}-1\right)}{2}$$

double points, that is: our curve has $\frac{n}{2}$ $(n - 2)$ *real double points*; all of them have real branches.

2. We want to calculate the *class* k of the curve. Evidently, k is equal to n, because there are two tangents to each ellipse, and there are $\frac{n}{2}$ ellipses. If we revert to the general formula, then we have $k = n(n-1) - R$; R is the class reduction, due to the existence of the double points. So R can be calculated by

$$R = n(n-2),$$

that is, the class of our curve is reduced by $n(n-2)$ units with respect to the class of the non-singular curve, *so that again each double point corresponds to a reduction of two units.*

3. We verify the existence of $\frac{n}{2}(n-2)(n^2-9)$ bitangents of the curve. We have

a) for every two ellipses four common tangents, which have to be seen as bitangents for the total curve. This yields in total

$$\frac{\frac{n}{2}\left(\frac{n}{2}-1\right)\cdot 4}{2} = \frac{n\,(n-2)}{2}$$

bitangents;

b) from each line connecting two intersection points of two ellipses (not necessarily of the same pair) we have four bitangents. This yields:

$$\frac{4\frac{n}{2}\,(n-2)\cdot\left(\frac{n}{2}\,(n-2)-1\right)}{2}$$

bitangents;

c) finally, a third type of bitangents is added, for the reason that from each intersection point of two ellipses it is possible to have two tangents to each of the remaining $\frac{n}{2}-2$ points. These tangents are counted twice as bitangents of the whole curve, as can be shown again in many ways. We have $\frac{n}{2}(n\text{ -}2)$ intersection **[195]**
points for each pair of ellipses, and for each of them $\frac{n-4}{2}\cdot 2$ tangents of the kind mentioned, all of which are real. Therefore this yields $\frac{n}{2}\,(n-2)\cdot\left(\frac{n}{2}-2\right)\cdot 2\cdot 2$ bitangents of the whole curve.

Adding the three kinds of bitangents we obtain the totality of bitangents:

$$\frac{n}{2}\,(n-2)\left(1+n^2-2n+2n-2-8\right)=\frac{n}{2}\,(n-2)\left(n^2-9\right).$$

So we have verified in our special curve the existence of all the $\frac{n}{2}(n-2)(n^2-9)$ bitangents of the general curve. Of these bitangents, sometimes four are brought to coincide, sometimes two, but they are all real with real contact points.

4. *Inflection points* of course do *not exist* in our degenerate curve.

Now we move from this degenerate curve of degree n, which consists of $\frac{n}{2}$ superposed ellipses $\Omega_1=0,\ldots,\Omega_{\frac{n}{2}}=0$ and that therefore has the equation

$$\Omega_1\Omega_2\ldots\Omega_{\frac{n}{2}}=0,$$

over to a *neighbouring curve*

$$\Omega_1\Omega_2\ldots\Omega_{\frac{n}{2}}=\pm\varepsilon.$$

The $\frac{n}{2}(n\text{ -}2)$ double points of the degenerate curve here yield $n(n-2)$ real inflections.

The bitangents, however, can be partially real and partially imaginary; this depends on the kind of transition to the neighbouring curve, that is, from the choice of the upper or the lower sign of ε. But surely no isolated bitangents will arise, because the contact points of the bitangents – as long as the bitangents are real – are also real due to the shape of the initial figure.

So we have reached our aim; we have

$$w'=n(n-2),t''=0$$

and therefore

$$w'+2t''=n(n-2),\ \text{as was to be proved.}$$

The example for an arbitrary even n was a generalization of the considerations made for the case $n=4$, with the difference that, in counting the bitangents, a new kind c) arose.

The Same for Odd n

Now I leave this case and consider an *odd* n; here, too, I start with the lowest case that presents a situation which is typical in the general case, namely $n=5$.

In accordance with the previous considerations, we construct at first a degenerate **[196]**
C_5, in which the various relations can be observed clearly, then passing from here to a general C_5.

The relationship that we should find is

$$w' + 2t'' = 15.$$

Our special C_5 is composed of a C_3, whose equation is $\varphi = 0$, and of an ellipse $\Omega = 0$, so that its equation is

$$\varphi \cdot \Omega = 0$$

The question is, how do we need to choose the C_3 and the C_2 in order to obtain suitable conditions for our purposes.

Figure 121

I already highlighted that a curve C_3 always has an odd branch, which we want to draw in particular in such a way that it presents only one asymptote (Fig. 121). We can construct this branch so that its shape is symmetrical with respect to a point of the asymptote, so that an inflection point falls at the intersection O of the asymptote with the curve, while the two others lie equidistantly on the straight line through O on opposite sides. Through dilatations, that is with an appropriate affine transformation, we obtain a shape particularly apt to our purposes, like the one of Fig. 122. The matter for us is not of having an overlook on all the possible shape of a C_3, but to distinguish a C_3, which is particularly appropriate for our purposes.

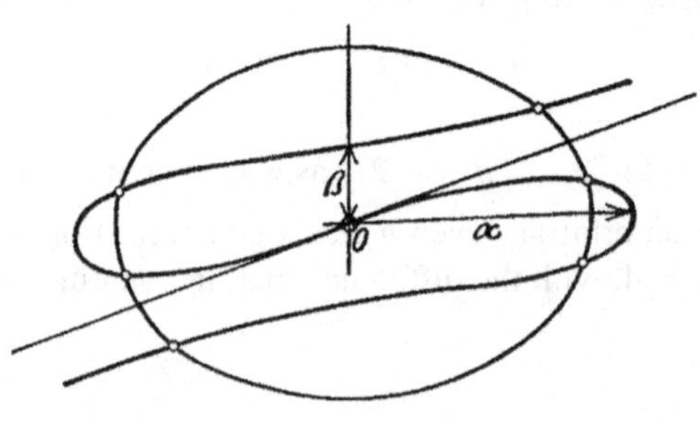

Figure 122

If we call β the minimum distance from the point O to the flanks of the curve stripe, and α the maximal distance to its vertices (Fig. 122) and we take two quantities a and b so that

$$a < \alpha, \quad b > \beta, \quad \text{while } a > b$$

we obtain, when choosing a and b as semi-axes of the ellipse centred in O, all six the possible intersections of the C_3 with the ellipse, guaranteed by Bézout's theorem, as real points. This continues to hold even when rotating the ellipse around O, that is, whatever azimuth we take for the axes. On the other hand, two such ellipses cut each other in four more real points, so that we have the possibility to construct in **[197]** a convenient way also higher curves $C_7, C_9, \ldots$, which I am not going to develop particularly in the following.

Let us keep studying our curve C_5, for which I can show things very concretely. First of all, the *class*.

The class of the C_3 is evidently equal to $3 \cdot 2 = 6$, that of a C_2 equal to $2 \cdot 1 = 2$, therefore the class of our curve is $6 + 2 = 8$. Being, due to the general theory, the class $k_5 = n(n-1) = 5 \cdot 4 = 20$, we have a reduction by 12 units, so that each double point produces again a reduction by two units.

The number of real inflection points is 3, because the C_3 has three, while the ellipse has no inflection points.

The following step is now to verify that our curve has $\frac{n}{2}(n\text{-}2)(n^2-9) = 120$ bitangents. These are, first of all, the tangents in common between the curves C_3 and C_2. This yields, since C_3 is of the sixth class and C_2 of the second, 12 bitangents of the C_5. Are they real or imaginary? Surely not all of them are real, as we can understand looking at Fig. 122. Without further justification we can claim: *Those of our 12 tangents common to the curves C_3 and C_2 which are real, also have real contact points, and not complex conjugate ones, so that they are not isolated bitangents.* This emerges from the fact that the contact points of such common tangents, which should be distributed along the curves C_3 and C_2, satisfy separate algebraic conditions.

From the double points of our curve, more tangents can be conducted: though not to the ellipse, but to the C_3, which touch the curve in a point different from the double point considered. There are four tangents each. The number 4 arises from the fact that from an arbitrary point, not on the C_3, six tangents to the latter can be conducted, but if the point is about to reach the C_3, two of these six tangents come to coincide at its proper tangent, so that there remain only $6 - 2 = 4$. This yields $4 \cdot 6$ total tangents (from each of the six double points four bitangents). When we untangle the double points we obtain $4 \cdot 6 \cdot 2 = 48$ bitangents, because each of our tangents (like in the case of an even n) yields two bitangents of the general C_5. Again, we do not know if these, taken individually, are real or imaginary, and we will leave this unexplored. But one thing is certain again:

These 48 bitangents, which we see as separate for the neighbouring curve $\varphi\Omega = \pm\varepsilon$, are surely not isolated bitangents, since they are real. Because, when going back to the degenerate curve, one of the contact points must become a double point, the other one must become distant from it, while for an isolated bitangent this passage should bring both contact points together at the double point.

The last category of bitangents is represented by the straight lines connecting the **[198]** double points to one another. There are $\frac{6 \cdot 5}{2} = 15$ connecting lines. Each of them has, however, to be counted with multiplicity four, as in the case of an even n, so that we get the proposition: *The fifteen lines connecting the six double points yield* $15 \cdot 4 = 60$ *bitangents for the neighbouring curve.*

We continue straightaway: *As far as the arising bitangents are real, again they are not isolated, because all the contact points have to be searched in the neighbourhood of the two double points, which we initially connected with a straight line.*

Counting the bitangents $(12 + 48 + 60)$ in this way the existence of all the 120 bitangents of the general theory has been verified in our curve.

Now we have all that is necessary to move to the *neighbouring curve* $\varphi\Omega = \pm\varepsilon$. As real inflections we clearly have the three inflections of the C_3 and two for each of the six double points, that is $3 + 2 \cdot 6 = 15$. Since these are *real bitangents*, they are not isolated, therefore $t'' = 0$. So, here, too, we correctly have: $w' + 2t'' = 15 = 5 \cdot 3 = n(n-2)$.

The same can be done for an arbitrary odd n. I omit the development that starts from a degenerate C_n, composed of a curve C_3 and of $\frac{n-3}{2}$ mutually congruent real concentric ellipses, whose principal axes follow each other with an inclination of $\frac{2\pi}{n-3}$.

In any case, also for an arbitrary odd n we can obtain examples of non-singular curves, for which it is

$$w' + 2t'' = n(n-2).$$

Explanation of the Continuity Proof Referring to Examples. Performing the Proof

After having constructed these examples, we go back to the *general reflection* about the proof to be performed. We know that we can always organise the passage from the representational point 1 of a non-singular C_n to the representational point 2 of a different non-singular C_n in such a way that we cross the manifold $D = 0$ only at individual points that correspond to a C_n without other singularities than a single double point, which reduces the class by two units. Now we want to know, what happens with the relationship $w' + 2t'' = n(n-2)$ existing for 1 when we pass to 2. We need to consider what can get lost or be gained.

One could think, initially, that a loss or gain of inflections w' or bitangents t'' occurs when one crosses a curve with an ordinary double point. But this is not
[199] the case, rather the following proposition holds: *The number* $w' + 2t''$ *remains unchanged when we cross the manifold* $D = 0$ *as convened before.*

Of course, we need to consider here that the ordinary double point of the C_n may be a double point with real branches or an isolated double point (Fig. 123).

In the first case it is clear that two real inflection points of the curve get lost when passing to the curve with one double point, but they appear again as soon as we untangle the double point, *so that* w' *remains unchanged.*

In the second case no real inflections are at all involved in the limit passage, so that a change of w' is not in question from the beginning.

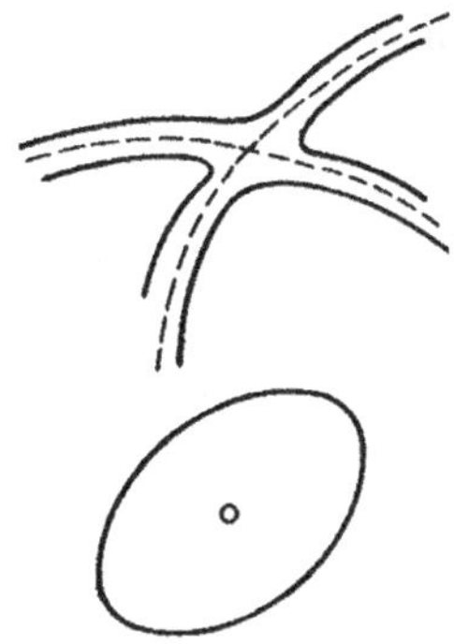

Figure 123

Thus the first step is done:

When going across a curve with a double point, which reduces the class by two units, the number w' of the inflections of the non-singular curve remains unchanged.

But, exactly in the same way, t'' also remains unchanged. Indeed, generally speaking, the number of the real bitangents changes, but the concerned bitangents are not isolated ones. One can clarify this upon looking at Fig. 124, where four bitangents become imaginary – in the transition of one of the ovals to an isolated point that becomes subsequently imaginary – but it is evident that they might neither have been nor might become real isolated bitangents.

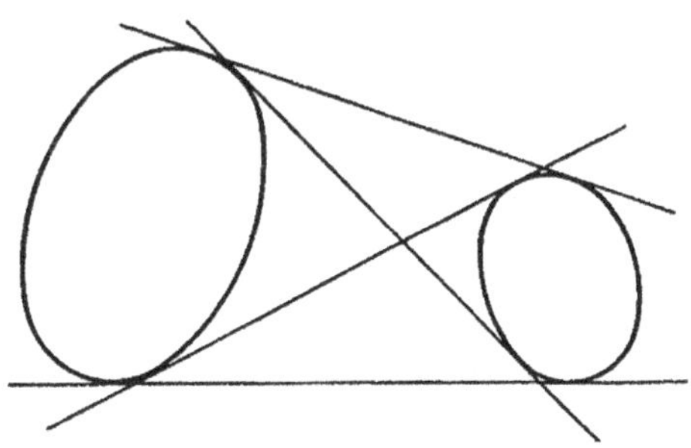

Figure 124

The same happens in the case of real bitangents, which can eventually be absorbed by a double point with real branches. Therefore:

The crossing of the curve with a double point can result in a great change in the number of the real bitangents of the curve, but the number of the real isolated bitangents remains unchanged.

If we sum up what we said about w' and t'', we see *that in the considered transition the sum $w' + 2t''$ remains, indeed, unchanged – before and after.*

With this we have completed studying the passage through the surface $D = 0$. Of course the matter can become more complicated if we cross the surface $D = 0$ in points corresponding to higher order singularities. But the beautiful thing is that

[200] we can completely avoid the analysis of these higher order cases. If one made the effort to study in depth the influence of crossing higher order singularities of the surface $D = 0$, there is no doubt that he would find as a final result that the sum $w' + 2t''$ would remain unchanged after the transition over the singularity. This is the case because it is always possible to get from the initial point to the endpoint of the path, bypassing the higher order singularities of $D = 0$.

Thus, if the sum $w' + 2t''$ remains the same when we cross the surface $D = 0$, how can it then change at all?

We will have to study the points, in which this can happen, only under the condition $D \gtrless 0$. If, namely, at the same time it were $D = 0$, then we would be confronted with a "higher occurrence", which we can always avoid thanks to our general reflections. Therefore, the following precise question is posed to us:

How can the number $w' + 2t''$ change when passing from a representational point to the other, if we are allowed to consider only situations that can be expressed by means of a single algebraic equation in the coefficients of the C_n, and in particular if we leave aside the case $D = 0$ because it has already been treated?

Since the total number of inflection points of a non-singular curve C_n is determined by n, the number w' can clearly change only if two real inflection points on the non-singular curve come to coincide and then become imaginary, or vice-versa.

On the other hand, correspondingly, t'' can change only if two isolated bitangents come to coincide and become then imaginary, or if an isolated real bitangent is transformed into a non isolated bitangent, or vice-versa.

We do not even need to consider the first of the two possibilities mentioned for the change of the number t'', because the requirement contains two conditions. Indeed, if two isolated bitangents come to coincide, we can have a quadruple tangent[166], or a tangent that touches the curve at two points each with multiplicity two (because the two conjugated complex contact points must behave in the same way).

The only change of t'', which can become relevant, therefore consists in the fact that from an isolated bitangent a non-isolated bitangent arises, or vice-versa.

And now things luckily go in such a way that the two changes of w' and t'', which are the only still to be studied, occur necessarily simultaneously in a manner that leaves $w' + 2t''$ unchanged.

We will explain the proposition that we have established here first of all through drawn curves, and then ask:

[201] *Is it possible to transform what we see here into exact theorems about idealised algebraic curves?*

We trace a branch of a curve with two real inflections, let them move close together until they coincide and then let them become imaginary. In Fig. 125a we clearly have a bitangent with two real contact points, from this we obtain in Fig. 125b a so-called four-point tangent and in Fig. 125c, in algebraic terms, an isolated bitangent, because a real bitangent can become imaginary only if it comes

[166] Translator's note: A quadruple tangent is a tangent that touches the curve in four distinct (imaginary) points, while a four-point tangent a tangent that touches the curve in one point with multiplicity four (see later).

to coincide with another real bitangent; but this does not occur in the present case. Therefore:

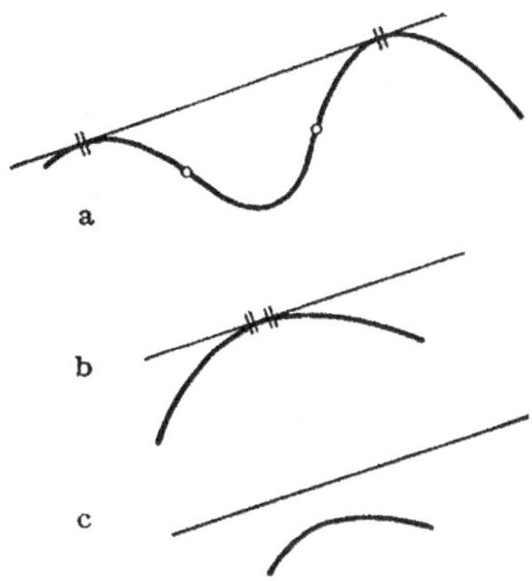

Figure 125

While the two inflection points come to coincide, becoming then imaginary, also the contact points of the bitangents move together and become thereafter imaginary. The bitangent itself remains real in this passage.

So, if w' is reduced by two units, t'' increases by one, so that, since t'' is multiplied by two in the sum $w' + 2t''$, the relation remains correct.

I want to explain this further with the two shapes of the C_4 which we had produced before.

We had started with two mutually intersecting ellipses and constructed first the curve $\Omega_1 \Omega_2 = -\varepsilon$, which consists of four separated ovals. Now I draw your attention to the fact that each individual oval has a bitangent (Fig. 126). If we now let ε increase, these bitangents may be pushed away; in this case we obtain a figure as shown in Fig. 127. Here there no longer exist real inflections, instead there are four isolated bitangents, that make the relation $w' + 2t'' = 8$ still hold.

In the second case, we put $\Omega_1 \Omega_2 = +\varepsilon$, and obtained a belt curve, with eight real inflections at first (Fig. 128). The outer branch has four bitangents; if ε increases **[202]** there arrives a moment at which the four bitangents are transformed into four four-point tangents and then into four real isolated bitangents, which enclose the branch, that now has an oval shape, without meeting it (Fig. 129). But in the meantime the eight inflections of the curve branch have disappeared. Therefore, here too the relation $w' + 2''t = 8$ remains unchanged. The question is: Why is it so, and not only in the particular case $n = 4$ but always, for any n? Here I answer, loyal to the principle of the present lecture course: *Firstly, the figures that we obtain for an arbitrary n are very similar to those observed now in our example. But the problem is to examine why is it that from these figures exact conclusions about idealized algebraic curves can be derived.*

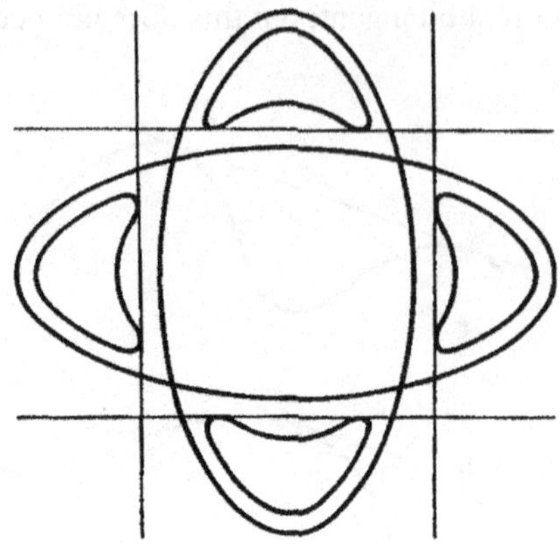

Figure 126

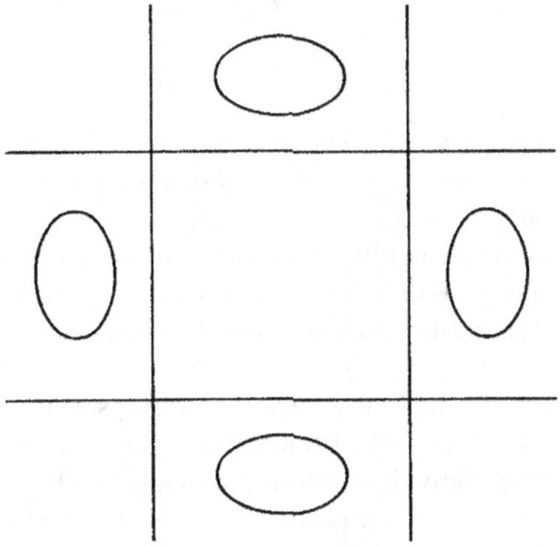

Figure 127

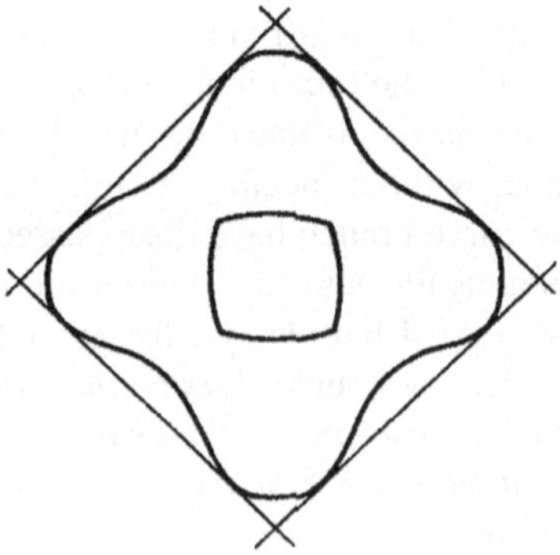

Figure 128

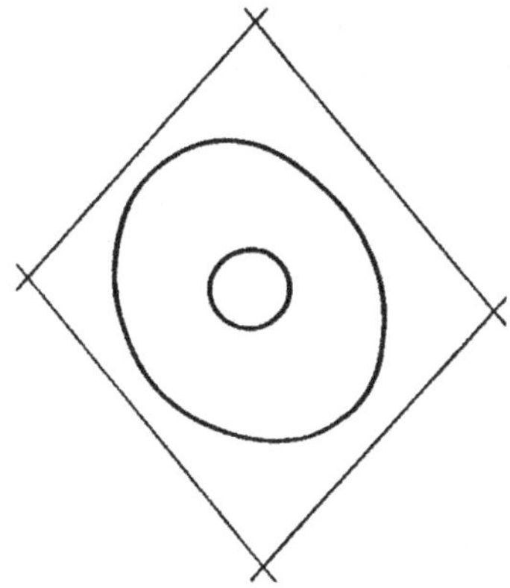

Figure 129

I sketch this as follows: The matter will be initially to consider certain empirical figures. Some branch curve that does not go to infinity may have a bitangent whose contact points follow one another after passing through a finite piece of curve. One glance at Fig. 130 shows that between the contact points there are at least two in- **[203]**
flection points.

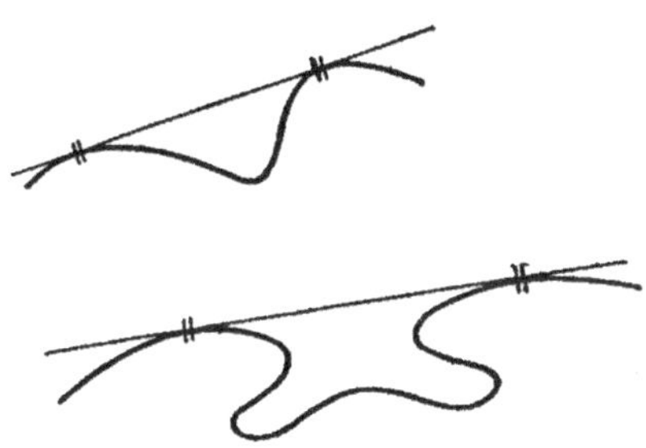

Figure 130

Indeed, their number can be equal to 2. Conversely, if we provide a branch of a curve, which originally had no inflections, with two or more inflection points by pushing in its parts, we will obtain al least one more double tangent.

Secondly, we will have to transpose these theorems to regular idealised curves, where each time the decisive point of the proof will be given by the correct use of Rolle's theorem or of Weierstraß's existence theorem for the maximum of a continuous function. If the curve branch has a bitangent, as before, we can choose the x-axis parallel to the bitangent and think of the part of the curve between the two contact points as described by a suitable parameter t in the form $x = \varphi(t), y = \psi(t)$. At the contact points it is then $\frac{dy}{dx} = \frac{\psi'}{\varphi'}$ equal to zero. But y needs to have a maximum and a minimum in the interval. There is once again $\frac{dy}{dx} = \frac{\psi'}{\varphi'}$ equal to zero. We deduce that $\frac{d^2y}{dx^2}$ vanishes at least twice in the interval, and this is what we wanted to prove. In the corresponding manner we find the inverse theorem.

Thirdly, someone will say that – when considering only "general" cases relative to the kind of algebraic curves – there are also *only* two inflections between the contact points of a bitangent, which has to come loose from the curve when this varies slowly in shape. Indeed, if there were more than two inflections, one could deform the branch in such a way that the two contact points of the bitangents move close together and this would then give rise to a case of higher occurrence, which we can avoid thanks to our general reflections.

Therefore, the three issues considered together yield that for the idealised algebraic curve the sum $w' + 2t''$ is constant with respect to the changes that we want to study and, therefore, that *this sum always takes on the constant value* $n(n-2)$ *for the curves* C_n free from higher order singularities, *q. e. d.*[167]

[204] The proof thus completed of our theorem differs from the one presented in volume 10 of *Mathematische Annalen* only for the fact that at that time I referred more directly to the empirical figures, while here I entered into further detail about the transposition of the conclusions from the figures to the idealised algebraic curves.

---

[167] [In Klein's paper cited on p. [178] a more general relation is also proved

$$n + w' + 2t'' = k + r' + 2d'',$$

where n, w', t', k, d'' have the meaning known to us and r' is the number if real cusps. How higher singularities can be included in this relation has been shown by *Alexander Brill* (*Mathematische Annalen*,Vol. 16 (1879), pp. 348–408); his study is purely algebraic, different from the one presented above. See also the proof by Christian Juel (*Mathematische Annalen,* Vol. 61, 1905, and his report of the sixth Scandinavian congress of Mathematicians in Copenhagen (1925), pp. 119–126). The extension to complex curves has been given by *Frederik Schuh* in the paper: On an expression for the class of an algebraic plane curve with higher singularities (*Academy of Sciences Amsterdam* 1904, pp. 42–45). If Σu is the sum of the degrees of all singularities with real point, Σv the sum of the classes of all singularities with real tangent, then it holds, according to *Schuh*:

$$n + \Sigma v = k + \Sigma u.$$

A very recent paper that extends Klein's research is the one by *Temple Rice Hollcroft,* Reality of singularities of plane curves, *Mathematische Annalen* 97 (1927), pp. 775–787, in which the author discusses the maximum number of real cusps of an algebraic curve of given degree.]

Third Part: About the Perception of Idealised Structures by Means of Drawings and Models [205]

I. About the Perception of Idealised Structures by Means of Drawings and Models

A main theme within the topics treated in this lecture course has been the distinction between empirical space intuition, with its limited precision, and the idealised conceptions of precision geometry. As soon as one becomes aware of this difference, one can choose his way unilaterally in one or in the other direction. One possibility could be to renounce rigorous definitions and undertake to construct a geometry only based on the evidence of empirical space intuition; in this case one should not speak of lines and points, but always only of "stains" and stripes. The other possibility is to completely leave aside space intuition since it is misleading and to operate only with the abstract relations of pure analysis. *Both possibilities seem to be equally unfruitful*: In any case, I myself always advocated *the need to maintain a connection between the two directions, once their differences are clear in one's mind.*

A wonderful stimulus seems to lay in such a connection. This is why I have always fought in favour of clarifying abstract relations also by reference to empirical models; this is the idea that gave rise to our *collection of models* in Göttingen. I want to use the last hours of this lecture course presenting to you a series of interesting models and their related explanations.

Gestalt Relations of Non-Singular Space Curves, in Particular C_3 (Projections of the Curve and Plane Sections of its Tangent Surface)

I am starting with geometry of space, and specifically with *space curves*.

In precision geometry space curves are defined as regular idealised curves by means of three equations:

$$x = \varphi(t), \quad y = \psi(t), \quad z = \chi(t),$$

where t varies in its interval. φ, ψ, χ need to be continuous functions, not all of them constant, twice differentiable and whose second derivatives are, furthermore, piecewise monotonous.

For such curves one can speak of a "tangent" and an "osculating plane".

F. Klein, *Elementary Mathematics from a Higher Standpoint*,
DOI 10.1007/978-3-662-49439-4_11

[206] One learns that the equation of the tangent at the point $x_0 = \varphi(t_0), y_0 = \psi(t_0), z_0 = \chi(t_0)$ is given by

$$\left\| \begin{matrix} x - \varphi(t_0) & y - \psi(t_0) & z - \chi(t_0) \\ \varphi'(t_0) & \psi'(t_0) & \chi'(t_0) \end{matrix} \right\| = 0.$$

Of course, to make this "matrix" equal to zero means nothing else than imposing the system of equations

$$\frac{x - \varphi(t_0)}{\varphi'(t_0)} = \frac{y - \psi(t_0)}{\psi'(t_0)} = \frac{z - \chi(t_0)}{\chi'(t_0)}.$$

The equation of the osculating plane at the same point is

$$\left| \begin{matrix} x - \varphi(t_0) & y - \psi(t_0) & z - \chi(t_0) \\ \varphi'(t_0) & \psi'(t_0) & \chi'(t_0) \\ \varphi''(t_0) & \psi''(t_0) & \chi''(t_0) \end{matrix} \right| = 0.$$

Consequently, one has the propositions:

Connecting two points of a space curve with a secant and letting these points approach one another somehow, the secant has the tangent as its limit position.

Considering the plane passing through three points of a space curve and letting these points approach one other somehow, the plane has the osculating plane as its limit position.

The question is now what gestalt relations can be observed on a model of space curve in correspondence to these propositions.

I start by showing you a certain number of models of space curves, where of course a specific difficulty lies in the fact that I must limit myself simply to a *description* of the relations, inviting you to successively study by yourself these things on the model[168]. My lecture as such must take a dogmatic form.

At first, I will give a model of an ordinary space curve without a singular point. On it we choose an arbitrary point p and delimit in a neighbourhood of it a "sufficiently small" piece of curve k. Our main question will be: How is this piece of curve projected from any viewpoint O onto a projection plane, of course not containing the point O, situated arbitrarily? We get three different answers to this question, depending on whether the viewpoint O lies outside the osculating plane at p, on the osculating plane but not on the tangent at p, or finally on the tangent at the point p. Suppose that k' is the projection of k, and p' the projection of p.

[168] Translator's note: In the preface to the third edition, *Fritz Seyfarth* writes that, thanks to the addition of figures and explanations, the availability of the models is no longer essential. Of course, Klein's suggestion to verify the assertions on a 3D-model, potentially even with a geometric software, is still valid.

Then we have:

1. If O does not lie on the osculating plane of p, then k' is a piece of curve that has no singularities at p' (Fig. 131).

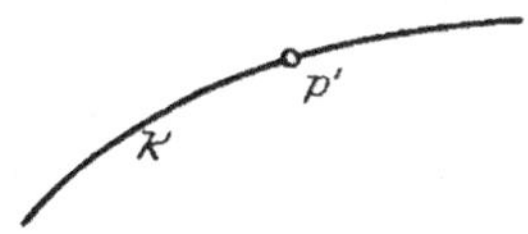

Figure 131

2. If O lies on the osculating plane at p, then the projection k' has at p' an **[207]** inflection point, because the space curve crosses the osculating plane (Fig. 132).

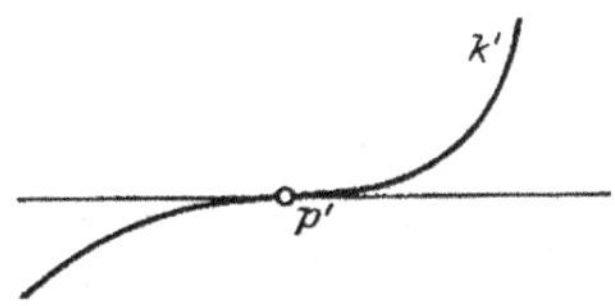

Figure 132

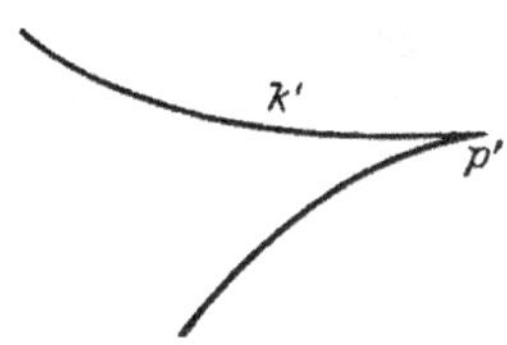

Figure 133

3. If O lies on the tangent at p, without being coincident with p, then k' has at p' a cusp (Fig. 133). This soon becomes intuitively clear if we choose the viewpoint not on the tangent at p but on a secant pp_1, and then move p_1 indefinitely toward p. We first obtain as the projection a curve with a double point. The two tangents at this double point correspond to the two tangents to the space curve at p and p_1, and they are surely skew for an appropriate choice of p_1. With the limit passage $p_1 \to p$ the two tangents at the double point come to coincide and the double point becomes a cusp.

These indications about the behaviour of the space curves can become considerably more complete if we now refer, in particular, to the *space curves of degree three*, because these – being algebraic – allow us to consider the whole behaviour of their projection and not only of a small piece of it.

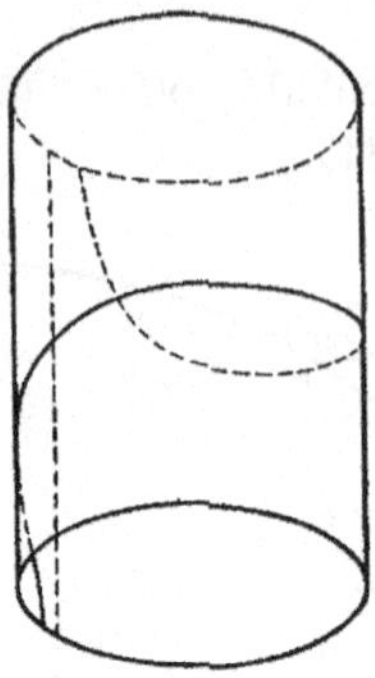

Figure 134

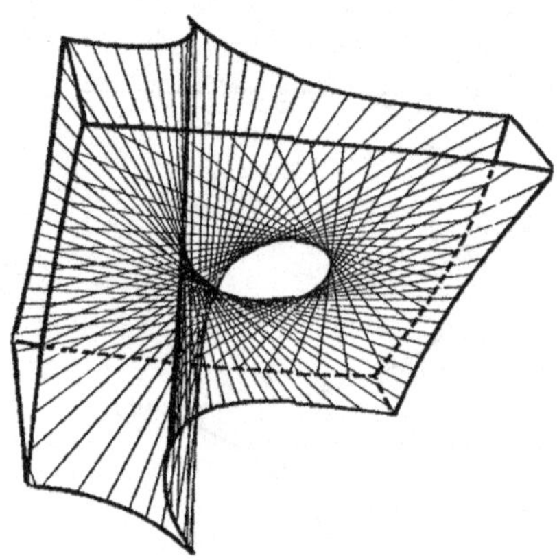

Figure 135

In Fig. 134 a so-called cubic ellipse is represented; it can be conceived as a section or better as a part of a section of a right elliptic cylinder with an oblique circular cone that has a generatrix in common with the cylinder. It is easy to understand that such a curve must be of the third degree, that is, a plane must intersect it at three points, two of which may be complex conjugated. Indeed, according to Bézout's theorem, two surfaces of the second degree intersect forming a space curve
[208] of degree four. If, now, the two surfaces have a common rectilinear generatrix, the space curve of the fourth degree must be composed of this generatrix and of a space curve of degree three. Fig. 135 shows the cubic ellipse and the developable surface formed by its tangents[169].

---

169 [The thread model of the tangent surface of the cubic ellipse, represented in Fig. 135, was built by *Walther Ludwig* and published by Martin Schilling, Leipzig. See the booklet accompanying the model: *Walther Ludwig*: Die Horopterkurve mit einer Einleitung in die Theorie der kubischen Raumkurve, again by Martin Schilling, Leipzig 1902.]

To better understand what follows we first state, without proof, the theorem: through any point of the space passes a straight line, which intersects our C_3 at two points. These two intersection points can be 1. real and distinct; 2. real and coincident; 3. complex conjugated and distinct. In the first case we have an ordinary secant, in the second case a tangent, in the third case a so-called "idealised" secant, which is nevertheless real, since it is the line connecting two complex conjugated points. If we now study the tangent surface formed by the tangents to the C_3, we see that it divides the space into two regions. In the first one, that, starting from the space curve, extends between the two wings of its tangent surface, lie the points through which an "idealised" secant to the C_3 passes, in the second one lie the points on ordinary secants, and on the surface itself there are the points through which the tangents to the curve pass. If we now project the C_3, from an arbitrary point, which is not on its tangent surface, the rays of projection form a cone of degree three with a double edge. The projection cone is of degree three because the C_3 is intersected at three points by any plane, and consequently any plane passing through the centre of projection intersects the cone of projection at three straight lines. The double edge is identical to the secant to the curve passing through the centre of projection, it is therefore always real, but it is isolated if the centre is in the first region. If we now cut the cone of projection with a plane, we obtain as section a plane curve of the third degree with an ordinary double point and an inflection point – if the cone has an ordinary double edge. To the inflection point obviously corresponds a point on the space curve, whose osculating plane passes through the centre of projection (Fig. 136a). But if we place the viewpoint in the region of the "ideal" secants, then we obtain as section a plane curve of the third degree with an isolated double point and three real inflection points. Indeed, in this case there are three points on the space curve whose osculating plane passes through the viewpoint (Fig. 136b). – The passage from Fig. 136a to Fig. 136b occurs when the viewpoint moves along the tangent surface, but not on the C_3 itself. The two tangent planes to the cone of projection, which touch along the double edge, then coincide in a double plane. The curve obtained by projection acquires a cusp, while still maintaining its inflection **[209]**
point. The cusp corresponds to that point of the space curve whose tangent passes through the viewpoint (Fig. 136c).

Up to now things are not difficult, because the three figures drawn simply follow each other in a continuous way. Things become more difficult when we ask the new question: What is the shape of the projected figure if we place the viewpoint on the space curve itself?

It is clear, from the algebraic theory, that the projected figure has to degenerate into a conic section and a straight line, while the third degree cone of projection is decomposed into a cone of the second degree and the osculating plane through the viewpoint. But how can this separation of the straight line from the curve occur despite the continuity? I maintain that Fig. 136c is transformed into Fig. 136d, and from here into Fig. 136e. Of course, instead of Fig. 136c, 136d, and 136e there can be also figures corresponding to them collinearly[170].

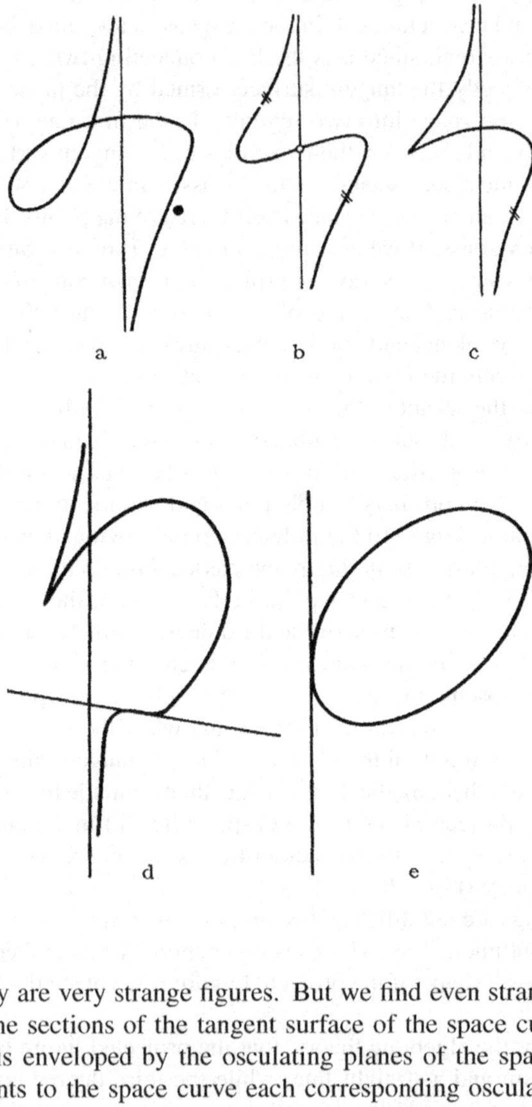

Figure 136

These already are very strange figures. But we find even stranger figures if we observe the plane sections of the tangent surface of the space curve. This developable surface is enveloped by the osculating planes of the space curve, that is, along the tangents to the space curve each corresponding osculating plane of the
[210] space curve is the tangent plane to the tangent surface. The points on the space curve determine in this way a regression or cuspidal edge of the surface. Along the space curve, the tangent surface is "as sharp as a razor's edge".

[170] Of course, the tangent at the inflection point (the trace of the osculating plane) shall not meet the cusp of the curve in 136d, because otherwise the curve would have 5 points in common with a straight line – the inflection point counting for three, the cusp for two.

It is wonderful that a sequence of tangents can form such a cutting edge of a surface, like the one you can see in the model.

The study of the plane sections of the tangent surface is, to speak generally, the dual counterpart to the study of the cone connecting the viewpoint to the space curve – thus anyone who knows projective geometry will immediately comment. Indeed, there is a general dual correspondence[171], which I am not going to deepen here, between:

a developable surface and a space curve,

a plane section of the surface and a projecting cone of the space curve,

a generatrix of the surface and a tangent to the space curve.

In particular for our curve of the third degree it holds – I will not go further on this – that the configuration generated by the points of the curve and their osculating plane is auto-dual.

Therefore, if we ask for the plane sections of the tangent surface, we only need to dualise the Figures 136a–e constructed before on p. [209]:

Fig. 136a is of class 4, because its double point reduces the class 6 of the general curve of the third degree by two units. The degree is of course 3, so that we can undoubtedly conclude:

An arbitrary plane section of the tangent surface is a curve of the fourth degree and of class three with a double tangent, corresponding to the double point of the projected curve (Fig. 137).

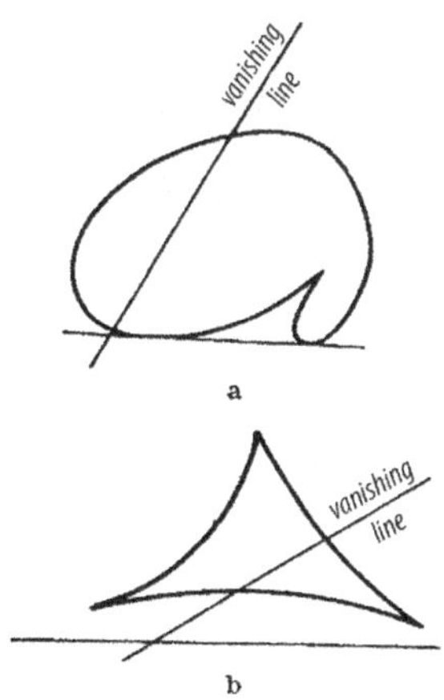

Figure 137

We know that a real double tangent can have real or imaginary contact points. So we obtain two possible shapes either in Fig. 137a a double tangent with real contact points, where the branch then has *one* cusp, or in Fig. 137b a double tangent with imaginary contact points, where then the real branch has *three* cusps. These figures correspond dually to the previous 136a and b.

[171] [See Vol. II, pp. [63]–[65]]

If we look at the model, aware of all this knowledge, we will not find – initially – sections of the tangent surface having the drawn shape. We first need to study the
[211] collinear transformations of the drawn figures. The matter is similar for conic sections, which can appear in three forms that are equivalent in the sense of projective geometry, but different to our eye (elliptic, hyperbolic, parabolic).

To obtain the "hyperbolic" type we trace a linear cut through the curves in our new Figures 137a and b and consider the secant line as a vanishing line of a suitable projection, by means of which we obtain two curves each with two asymptotes, along which the branches go to infinity[172].

Drawing the figures in an only qualitative way, I obtain the two configurations in fig. 138a and b, *and these actually appear when we cut the model of our tangent surface with suitable planes corresponding to the fact that the model with two wings extends towards infinity* (see Fig. 135).

The further question is: *What is the form of the section if the secant plane passes through a tangent to the space curve*?

It is immediately clear that in this case the generatrix of the tangent surface is detached from the curve of the fourth degree, leaving a curve of degree three and class three, namely a curve with a cusp (Fig. 138c). A deeper reflection shows that this curve touches our tangent at its contact point with the space curve.

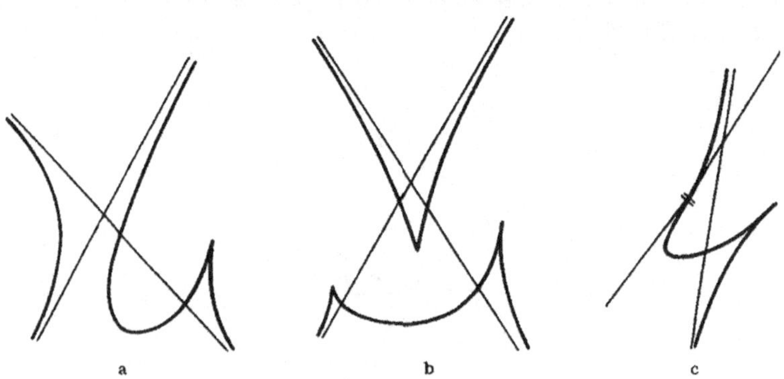

Figure 138

Such theorems can be easily established in an abstract way. But the question is how the transition from Fig. 138a to 138b passing through 138c can occur with continuity, in accordance with the transition of the curve of degree three with a double point to a curve with cusp. Thus I say, and please convince yourself looking at the model: We obtain Fig. 138c from a or from b if we let the two branches 1
[212] and 2 of Fig. 138a move close together in the way described by Fig. 139a or, if we

[172] [Fig. 137a and b illustrate themselves (without the added secant) the "elliptic" type (non-secant and non-touching vanishing line)].

let two cusps of Fig. 138b (see the following Fig. 139b), not belonging to the same curve branch, move close together. Similarly, we can move back from c to a and b by merging the inflection tangent of b with the related curve in one or the other way.

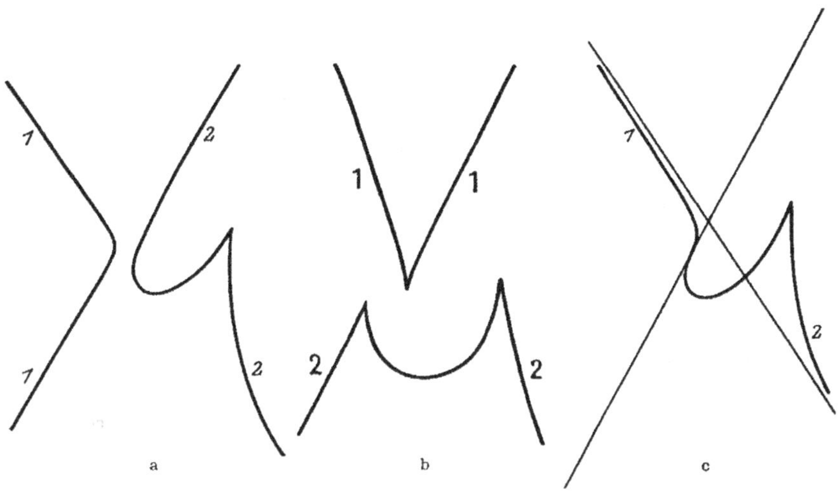

Figure 139

Finally we also consider the *section of our surface with an osculating plane.* We obtain a *conic section with a tangent counted twice* (Fig. 140). This figure results from Fig. 139c by inserting the intermediate Fig. 141. The latter evidently arises from Fig. 139c when – holding onto the inflection tangent – we bend down the infinite branch 1 while branch 2 is turned upward so that it presents a cusp pointing downwards. If the cusp and the inflection point are brought to coincide, then Fig. 140 arises.

Figure 140

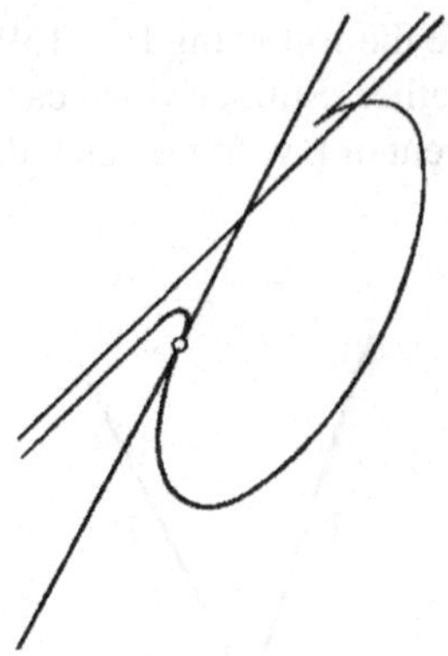

Figure 141

By the way, this is in the end the same transition that we have already studied on p. [209].

[213] After these considerations, I leave the request for each of you to verify them by yourself on the models, thus stimulating your intuition.

To the projective properties of the space curves now discussed, in particular of the space curves of the third degree, are naturally connected the studies concerning the *metric properties*. Here I am satisfied with the hints given concerning curvature, torsion[173] and tangent developable of a curve. In reference to the latter I remind you that, depending on whether one winds onto a plane a developable surface or a string onto a curve, the path is called an evolvent surface or an evolvent[174].

The Seven Kinds of Singular Points of Space Curves

I want to say something more about the *singular points of space curves*.

In the plane, we can make a scheme in four parts, characterising with the signs (+) and (−) the behaviour of the point (p) on the curve and of the tangent (t) to the curve depending on their moving forth or back when proceeding along the curve.[175]

[173] Translator's note: The torsion measures the deviation of the curve from the osculating plane. Klein has not used this term before.

[174] Among the textbooks, in which these things are presented thoroughly in a particularly clear manner we mention: *Georg Scheffers:* Einführung in die Theorie der Kurven in der Ebene und im Raume, Leipzig 1900. 3rd ed. 1923. [*Reinhold Lilienthal:* Vorlesungen über Differentialgeometrie 1. Leipzig 1908.]

[175] Translator's note: Point p has the sign + or − when it moves resp. to the right or to the left. The tangent proceeds clockwise on the osculating circle and its sign is + or − if its movement is resp. concordant or not with that of p (being linked to the concavity). See Fig. 143.

So we obtain, besides the ordinary point, the three kinds of singular points, different in their shape as follows (Fig. 142):

p	t	
+	+	ordinary point (1)
+	−	inflection point (2)
−	+	ordinary cusp (3)
−	−	ramphoid cusp (4)

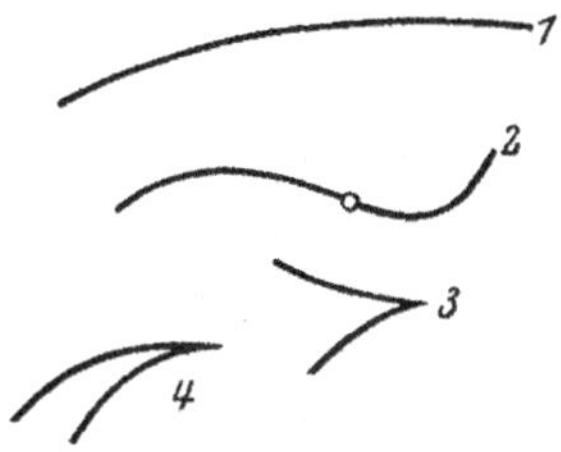

Figure 142

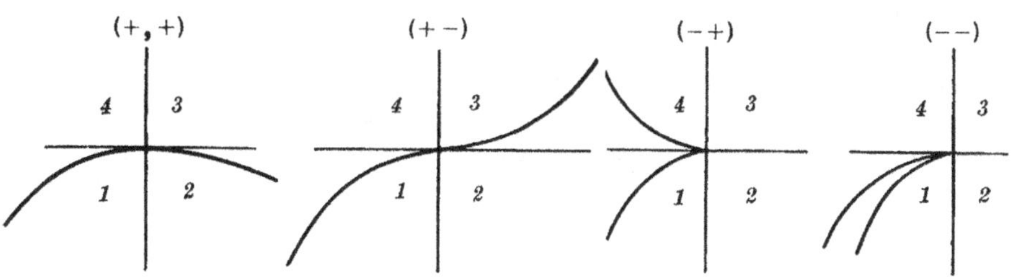

Figure 143

The corresponding scheme for three dimensions was given in 1847 by *Christian von Staudt* in his "Geometrie der Lage" (p. 113). Here one needs to consider, besides the point (p) on the curve and the tangent (t), also the osculating plane (e) and to form all the eight combinations of the signs (+) resp. (−) three by three.

p	t	e
+	+	+
+	+	−
+	−	+
−	+	+
+	−	−
−	+	−
−	−	+
−	−	−

[214] *Thus, from the simple point of view of their shape, one has to distinguish seven cases of singularities among the space curves*, for which no particular denomination has been created, however. In any case, it is clear how they can be characterised. The eight different cases are described through models, which were constructed by *Christian Wiener*[176].

In order to make the understanding easier, I will go back for a moment to the plane curves. If we choose the point considered as the origin and the tangent as the x-axis of an orthogonal coordinate system, the four separate cases identified above for the plane curve can be obviously distinguished by observing whether the curve, which comes from the first quadrant, continues into quadrant 2, 3, 4 or 1. See Fig. 143.

It is then natural to suppose, analogously, that the eight cases of the space curves can be described by the continuation of the curve into the eight different octants of the coordinate system. Indeed, this works. I lay this down in a more extensive manner with the following words: Construct a rectangular coordinate system, which has the tangent of the space curve considered as an axis, the corresponding osculating plane as a coordinate plane and the contact point as the origin. *Then, if the curve reaches octant 1, there are eight possibilities as to which of the eight octants the curve will continue into, and these eight shapes correspond to those counted above according to von Staudt.*

What happens precisely in the individual case has to be studied on the model, because a simple list offered through a table does not help the perception due to its uniformity. At the same time, the models make pertinent the question about the projection of the considered space curves from a changing viewpoint, the simplest case of which $(+,+,+)$ we already considered. The viewpoint will then be chosen:
[215] 1. outside the osculating plane or 2. on the osculating plane, or, finally, 3. on the tangent. What is odd here is that by projection strong singularities of the space curve can disappear from view. For instance, in the case $(-++)$ the projection from the tangent yields a plane curve $(++)$ without singularities, or - as one could say - with a "hidden singularity". A further odd property that I want to recall, is that the point $(---)$ of the curve yields a ramphoid cusp from which, taking it as viewpoint, even a neighbouring piece of the curve can be projected.

Moreover, I would like to suggest studying the related shape more deeply. In particular, the need comes to explore systematically the shape of the tangent surface, which snuggles close to the space curve in the eight cases, and to see in particular what kind of plane sections arise from it and how they are linked together by continuity.

For the analytical aspect, literature is available on the facts here exposed, which I want to present briefly. All the authors assume that x, y, z can be expanded in

[176] [See *Zeitschrift für Mathematik und Physik*, Vol. 25 (1880), pp. 95–97. The models can be obtained by Martin Schilling in Leipzig].

power series progressing according to entire exponents of t. The reference is to the following papers[177]:

Henry Burchard Fine: Über Singularitäten von Raumkurven. PhD thesis, Leipzig 1886; also the *American Journal of mathematics*, vol. 8 (1886), pp. 156–177;

Otto Staude: Über den Sinn der Windung in singulären Punkten von Raumkurven. Ibidem Vol. 17 (1895), pp. 359–380;

Alfred Meder: Über einige Arten singulärer Punkte von Raumkurven. *Journal für reine und angewandte Mathematik*, vol. 116 (1896), pp. 50–84 and pp. 247–264.

A discussion of the shape in the manner I am proposing should start from these publications.

With this I am concluding what I wanted to say about space curves and turn to the *surfaces*.

General Considerations on the Shape of Non-Singular Surfaces

If we consider all the classifications that we already had made for the plane curves (analytic, non-analytic, regular, etc.) we can certainly expect to find here also only exceptions if I limit my observations to *algebraic surfaces*, what indeed I am going to do for the sake of brevity

An algebraic surface is defined by an equation

$$f(x, y, z) = 0$$

where f is a polynomial.

It is usual to put the terms of f in order so as to group together the terms of equal degree:

$$0 = f_0 + f_1 + f_2 + \dots$$

Studying the surface next to one of its points, one conveniently choses this point **[216]**
as the origin of the coordinates, so that the constant term f_0 vanishes. So, generally speaking, one can assume the surface as given, in first approximation, by $f_1 = 0$, that is by the tangent plane at O. This equation can almost be understood as a definition of the tangent plane and one can say: *The tangent plane is defined by making the terms of degree one equal to zero.*

For a better approximation, one can put

$$0 = f_1 + f_2$$

and name this surface as an osculating surface of the second degree, etc.

177 [Also see pp. 255–272 of the textbook by Lilienthal mentioned on p. [213], and Meder, Analytische Untersuchung singulärer Punkte von Raumkurven, *Journal für reine und angewandte Mathematik*, vol. 137 (1910), pp. 83–144.

With this approach a subdivision of the points of the surface is immediately reached: If f_1 is not identically equal to zero, that is if a tangent plane can be determined, one speaks of an "ordinary" point of the surface; but if $f_1 = 0$ and $f_2 \neq 0$, then we have a double point; in general one speaks of an ν-ple point if it is $f_1 = f_2 = \ldots f_{\nu-1} = 0$, $f_\nu \neq 0$. In the following, we will sometimes use the expression *node* instead of double point.

What do we have to observe concerning the *ordinary points*?

First of all, we wonder at what curve the surface is intersected by the tangent plane at such a point. One assumes that the considered curve is given, in first approximation, by the simultaneous vanishing of the two equations $f_1 = 0$, $f_2 = 0$. So we compare the curve to the intersection of the tangent plane and of the osculating surface of the second degree. Since $f_2 = 0$ is a cone of the second degree, and $f_1 = 0$ a plane, this intersection evidently consists of two straight lines, which can be real or imaginary or that can coincide; and when transferring these distinctions to the shape of the intersection curve of surface and tangent plane, one has the following classification:

1. The straight lines are real: The intersection curve has a double point with real branches.
2. The straight lines are imaginary: The intersection curve has a double point with imaginary branches.
3. The straight lines coincide: The section curve has a cusp.

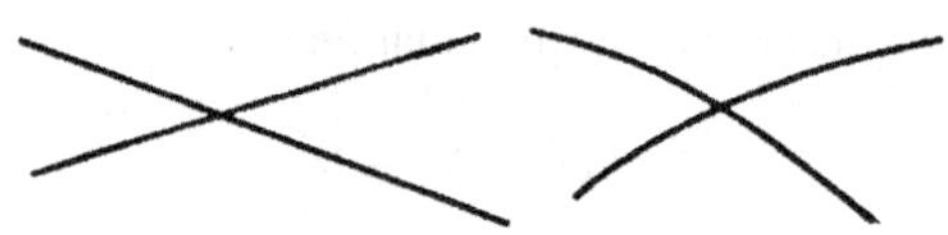

Figure 144

Figure 145

I repeat: *Apart from higher occurrences, where $f_2 = 0$, the tangent plane intersects the surface forming a curve with a double point, whose branches are real or imaginary or can coincide.* Consequently one speaks of "hyperbolic", "elliptic" and "parabolic" curvature of the surface.

[217] It is easy, by the way, here to think of various beautiful examples in which the curve considered, by intersection of the surface and the tangent plane, has higher order singularities, and to study the special case when a ramphoid appears as the section, or similar situations. If, in particular, it is $f_2 = 0$, the intersection of the tangent plane with the surface has a triple or higher point. This leads to the following observation:

As is known, in the elementary textbooks on differential and integral calculus the maxima and minima of functions in two variables are studied. This implies the study of the intersection of the surface $z = f(x, y)$ with the horizontal tangent plane.

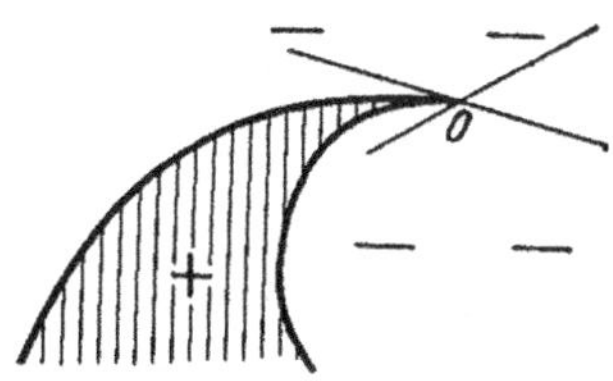

Figure 146

Suppose, for instance, that the intersection yields a ramphoid (Fig. 146), and that in the striped region the surface raises above the tangent plane, while sinking in the non-striped region; it is clear then, first of all, that in the contact point O there is neither a maximum nor a minimum of the function $z = f(x, y)$ represented by the surface.

But if we study the totality of the vertical planes passing through O, we see that each azimuth yields an intersection curve, which has at O a real maximum because, along these intersection curves, *z initially sinks constantly* starting from O. Indeed, it is impossible to *directly* reach the striped region from one of these curves starting from O. It is therefore not sufficient – in order to analyse the question whether a maximum or a minimum of a function z in two variables exists – to carry out the investigation only for this kind of sections. An extreme of the surface occurs only if *each* of the lines through O lying on the surface presents an extreme at O.

Using a purely analytical approach the matter seems anyway rather difficult to understand. Actually, all the textbooks have contained wrong statements to this regard, until for the first time *Giuseppe Peano* drew attention to the correct situation! By the way, also *Weierstraß* has always presented the correct theory in his lectures about calculus of variations.

Of course, we can also consider the metric relations of the surfaces. Particularly useful, for points for which $f_1 \neq 0$ and $f_2 \neq 0$, is here *Leonhard Euler*'s *general theory on curvature*:

One constructs the normal at the considered point, takes plane sections passing through it and constructs their circles of curvature. Then there are two mutually per- **[218]** pendicular normal sections, for which the curvature is a maximum or a minimum. I do not expose this in detail, but I mention that for the explanation of the usual theory of curvature in ordinary surface points models have been constructed again by *Christian Wiener*[178], that represent the circles of curvature in the individual normal sections.

[178] [Compare *Walter von Dyck*: Spezialkatalog der Mathematischen Ausstellung (Deutsche Unterrichtsausstellung in Chicago 1893), Berlin 1893, pp. 52.]

Finally I draw your attention to the models denoted in Walter von Dyck's catalogue[179] as "bean-shaped specimen", on which one should individually mark points of hyperbolic, elliptic or parabolic curvature. It turns out that our eye is here very uncertain, so that the definitions of the elementary theory of curvature already seem to have a very abstract character. One feels this particularly intensely when trying to perform the given task on such an intricate surface, given empirically, as for instance a bust. I recommend studying this.

About the Double Points of the F_3, in Particular its Biplanar and Uniplanar Points

Now I go over to the *discussion of the singular points of surfaces*, in the specific case of the surfaces of the third degree. According to our previous observation, the equation of a F_3 with a double point at the origin can be written as

$$0 = f_2 + f_3$$

We assume that $f_2 \neq 0$, so that we have an actual double point of the surface and not, for instance, a triple point.

Then it is usual to say that the cone of the second degree $f_2 = 0$ is a "first approximation" of the behaviour of the surface at the point O, and, depending on whether this cone is

a) an actual cone (real or imaginary) or

b) a cone that is decomposed into two (real or imaginary) planes, or finally

c) a cone that degenerates into a double plane,

we distinguish between

a) an *ordinary* double point, which is the cusp of a real cone or an isolated point

b) a *biplanar* point

c) a *uniplanar* point of the surface.

We call the always real intersection line of the two planes in case b) the *axis of the biplanar point.*

The question is: What is the shape of a surface of degree three that has a double point of the one or of the other kind?

[219] I observe, first of all: If we intersect the surface $f_2 + f_3 = 0$ with $f_2 = 0$, what remains is $f_3 = 0$, that is an equation representing a cone of the third degree. The two cones have, according to Bézout's theorem, $2 \cdot 3 = 6$ common generatrices: *Therefore, through a double point of a surface of degree three pass six straight lines of a surface*, where of course one has to consider that these straight lines may be real or imaginary, or they may also coincide.

Now I give some examples of F_3, in which the different kinds of double points occur, but I am not proving that things work similarly in the general case.

---

[179] [ibidem p. 54].

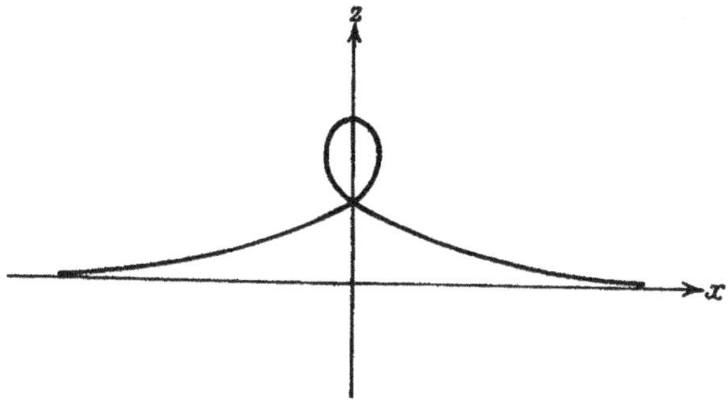

Figure 147

In order to obtain the surface with an *ordinary double point*, we limit ourselves to surfaces of revolution and draw, with a cut along the meridian (xz-plane), a curve of the third degree with a double point, symmetrical with respect to the z-axis, which has the x-axis as a so-called inflection asymptote[180] (Fig. 147). The shape of the figure is then clear. The six straight lines passing through the double point are here evidently imaginary. In an analogous way, we can construct a surface with an *isolated double point*. It is sufficient to draw a curve of degree three with an isolated double point as its section with the meridian (Fig. 148).

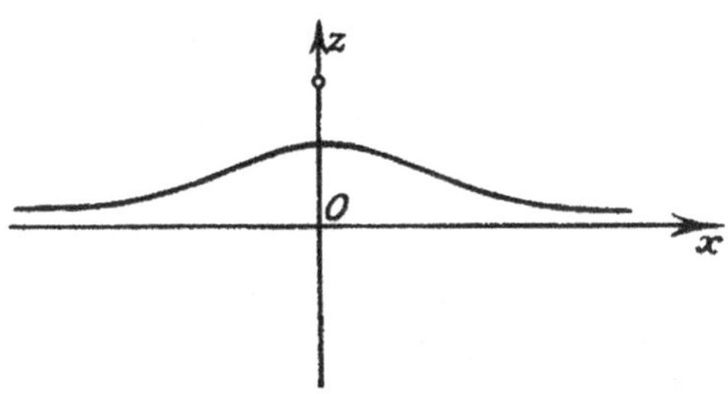

Figure 148

Further on, I mention a surface for which the six straight lines passing through a single double point are real. I choose a surface with four ordinary double points, which can be taken – in order to easily represent the surface analytically – as vertices of a coordinate tetrahedron. The equation of the surface that I have in mind is then, after having correctly chosen the multiplicative constants still at disposal due to the definition of the tetrahedral coordinates:

$$\sum_i \frac{1}{x_i} = \frac{1}{x_1} + \frac{1}{x_2} + \frac{1}{x_3} + \frac{1}{x_4} = 0.$$

[180] Translator's note: The point at infinity of the asymptote (and of the curve) is an inflection point.

[220] Multiplying the numerators by the denominators, we obtain the equation of degree three in x_1, x_2, x_3, x_4:

$$x_2x_3x_4 + x_1x_3x_4 + x_1x_2x_4 + x_1x_2x_3 = 0.$$

If we finally pass, by means of a projective transformation of the surface, from the tetrahedral coordinates to ordinary coordinates, thanks to the formulas

$$x_1 = x, \quad x_2 = y, \quad x_3 = z, \quad x_4 = 1,$$

the equation becomes

$$yz + xz + xy + xyz = 0.$$

One of the double points lies at O, the three others are infinitely far away.

In the equation thus described we have directly in front of us the two cones $f_2 = 0$, $f_3 = 0$ belonging to the origin. The cone $f_2 = 0$, which represents in a first approximation the surface in the neighbourhood of O,

$$yz + xz + xy = 0$$

obviously contains the three coordinate axes. The cone of degree three

$$f_3 = xyz = 0$$

is composed of the three coordinate planes passing through O, and therefore contains twice the coordinate axes, so that we can state:

The six generatrices, which in our surface pass though the node, coincide in pairs with the three coordinate axes.

Furthermore, you will easily prove by yourself that:

Along these generatrices the tangent plane of the surface coincides with that of the cone $yz + zx + xy = 0$, and is therefore constant.

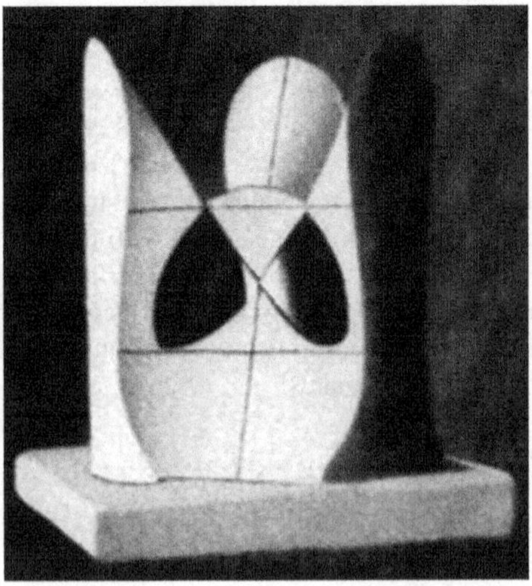

Figure 149

These statements hold, due to their projective character, also for the F_3 with four nodes lying in the finite. One should convince oneself of this by looking at the model that I am presenting (Fig. 149). It shows, first of all, the six edges of the double point tetrahedron as straight lines lying completely on the surface, along which the tangent planes run constantly; three more straight lines exist, separated from the former ones but still appearing as the section of the surface with a given horizontal plane (only one of these can be seen in the Figure), along which the tangent plane turns.

Now I pass over to the *biplanar points* of the F_3.

Now the understanding is no longer as simple; one can no longer get an im- **[221]**
mediate representation of the gestalt relations of a biplanar point, one has thus to consider the question concretely.

Let, in our equation

$$f_2 + f_3 = 0$$

f_2 be the product of two different linear factors; if we put $f_2 = 0$ we obtain two different planes. We take first of all the two planes as complex conjugates, that is we put for instance

$$f_2 = (x + iy)(x - iy),$$

so that the z-axis is the real intersection of the two planes, and therefore the axis of the biplanar point. The surface itself may then still have the simple equation

$$0 = x^2 + y^2 + z^3.$$

What does the biplanar point arising here look like? If we say that it is represented, in first approximation, by the two imaginary conjugated planes passing through it, this is not wrong given an appropriate interpretation, but not directly understandable. I rather express myself as following:

Our surface is a surface of revolution. We initially draw the meridian section with the xz-plane:

$$0 = x^2 + z^3.$$

This is a curve of the third degree with a cusp at O (Fig. 150). We obtain the surface itself by revolution of this curve around the z-axis[181]. In general:

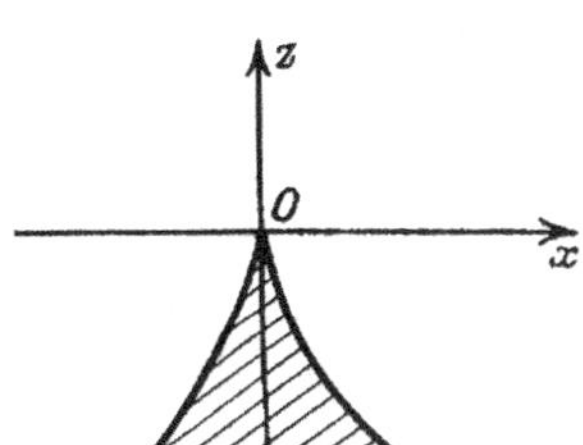

Figure 150

If a biplanar point has imaginary planes, then the surface has there the shape of a sharp thorn.

[181] The surface therefore represents an intermediate case between the two Figures shown on p. [219].

More difficult to understand is the case of a *biplanar point with real planes.* I take as example

$$0 = x^2 - y^2 + z^3,$$

where the planes of the biplanar point are given by

$$x + y = 0, \quad x - y = 0$$

that is, by two mutually perpendicular planes through the z-axis, which are inclined by 45°, resp. 135°, with respect to the xz-plane. The z-axis is again the axis of the biplanar point.

[222] If I cut the surface with one of these planes, what remains is

$$f_3 = 0, \quad \text{that is} \quad z^3 = 0.$$

Therefore: If the axis of the biplanar point is vertical, each of the planes intersects the surface forming a horizontal straight line counted three times.

If it is $f_3 = a_1 z^3 + a_2 z^2 x + \ldots$ we obtain, together with $y = \pm x$, an equation of the third degree in z/x. So we obtain: *Each of the two planes of the biplanar point intersects the surface in three straight lines, two of which may be complex conjugated.*

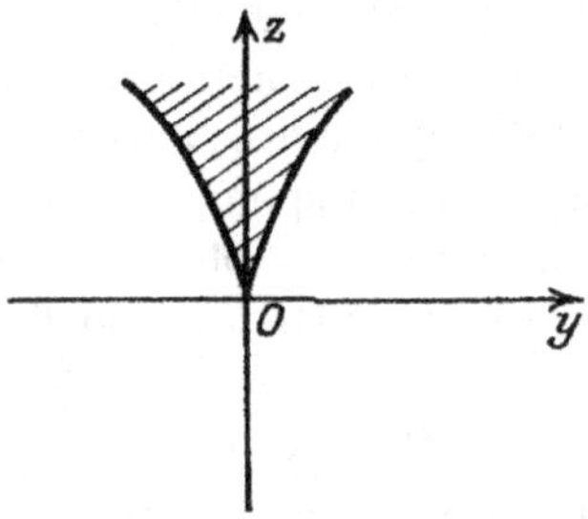

Figure 151

The intersections with the meridian planes passing through the z-axis are very strange. If, in the case $f_3 = z^3$ we perform the intersection with the plane $y = 0$, then it becomes $x^2 + z^3 = 0$, this means that the above considered curve has an upward cusp. Intersecting it with the plane $x = 0$, we obtain $-y^2 + z^3 = 0$, which corresponds to an analogous curve with a downward cusp (Fig. 151). So, while rotating the plane $y = 0$ onto the plane $x = 0$, the cusp reverts its direction; the transition occurs on the planes

$$y + x = 0 \quad \text{and} \quad y - x = 0,$$

whose section we have already studied. So we have the following proposition:

The meridian planes, which can be put through the vertically chosen axis of the biplanar point, intersect the F_3 generally at a curve with a cusp, so that the z-axis is tangent to a cusp. Only if the meridian planes are tangent planes of the biplanar point they intersect the F_3 forming three straight lines.

The family of the meridian planes is divided by the two planes of the biplanar point into two subfamilies. The matter is obviously thus that the intersection curves with the meridian planes of the one family have the cusp directed upwards, while the intersection curves with the meridian planes of the other family have the cusp directed downwards.

Figure 152 visualises the case of a biplanar point, for which three real straight lines lie on the two tangent planes; the meridian planes, which lie in the neighbourhood of these tangent planes, yield then intersection curves for which two more branches are added to the curve with a cusp.

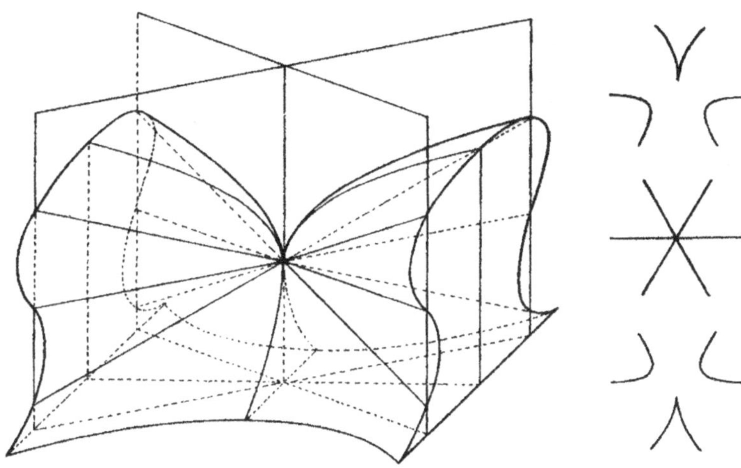

Figure 152

Just as it is not easy to imagine the shape of the biplanar point here discussed without further explanations, so it is also in the case of the *uniplanar point*[182].

Here we have as a "first approximation" a *double plane*, represented by the terms of degree two, let us simply say $z^2 = 0$. Then the equation of the surface is **[223]**

$$z^2 + f_3 = 0;$$

[182] These considerations about the shape of biplanar and uniplanar points seem to have been made for the first time by *Ernst Eduard Kummer* and *Ludwig Schläfli*; later on I used them in my paper "Über die Fächen dritter Ordnung" (*Mathematische Annalen*, vol. 6, 1873, pp. 551–81) [printed with additions by *Klein* and *Hermann Vermeil* in *Felix Klein:* Gesammelte mathematische Abhandlungen, vol. 2, pp. 11–62].

we choose f_3, for simplicity, so that only x, y appear in it; two cases are obtained, whose meaning becomes immediately clear. We put at first

$$f_3 = x(x^2 - y^2)$$

and then

$$f_3 = x(x^2 + y^2)$$

The section of the two cones $f_3 = 0, f_2 = 0$ yields, in the first case, three real straight lines, and in the second case one real and two imaginary straight lines (Fig. 153).

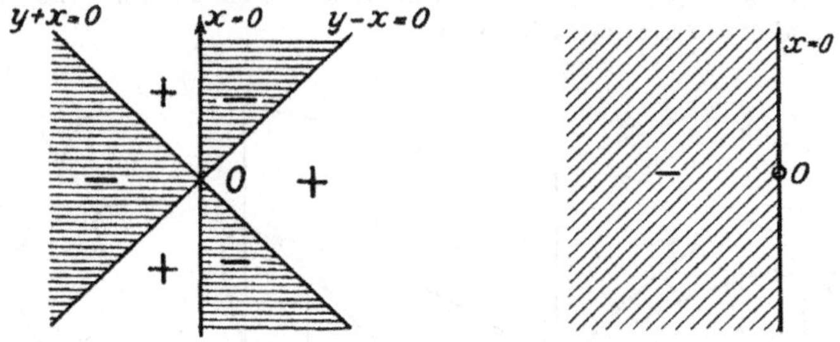

Figure 153

The general proposition is:

The plane of the uniplanar point cuts the F_3 in three straight lines passing through O, one or all the three of which are real.

[224] For z itself, the equation of the F_3 yields

$$z = \pm\sqrt{-x\,(x^2 - y^2)}$$

in the first case, and

$$z = \pm\sqrt{-x\,(x^2 + y^2)}$$

in the second case.

Except for $f_3 = 0, z$ becomes real only for negative f_3; indeed, in those regions that have negative f_3, there are for each pair of values x, y two real values of z. Consequently, the surface of the third degree covers doubly, seen from above, those parts of the xy-plane that are striped in Fig. 153, and it does not cover the other parts at all.

I explain this a little bit further:

If the three generatrices in the double plane of the uniplanar point are real, the F_3 extends itself inwards from three sides towards the uniplanar point, in such a way that – seen from above – it seems to cover doubly three sectors of the xy-plane. Every plane section through O, which does not coincide directly with the xy-plane, has a cusp at O.

In the second case, in which only one of the three generatrices lying on the double plane is real, the F_3 extends itself inwards from the left side of the x-axis towards the uniplanar point.

We may sum up saying:

The actual shape of the biplanar and uniplanar points of the algebraic surface becomes clear only if we consider not only the terms of degree two, but also those of degree three.

With our last reflections we were led so closely to the *shape of the surfaces of degree three*, that I want in the end to present these with more general details.

About the General Behaviour of the F_3

To provide, first of all, some historical information on the theory of the F_3, I mention that:

In 1849 *Arthur Cayley* and *George Salmon* proved that 27 straight lines exist on the F_3 (see *Cambridge and Dublin Mathematical Journal*, Vol. 4, 1849, pp. 118 and 252).

From then on their classification was studied deeply, in particular:

In 1863 *Ludwig Schläfli* discussed in detail the case in which the straight lines are real or imaginary and he also observed the singularities that can occur in a F_3 (see *Philosophical Transactions*, vol. 153, 1863, pp. 193–241).

In 1873 there are my already mentioned investigations about the gestalt relations of the F_3, in volume 6 of the *Mathematische Annalen*.

Thereafter:

In 1879, *Carl Rodenberg* in *Mathematische Annalen*, vol. 14, pp. 46–110, gave an analytic verification of my results found through considerations on geometric continuity. On the other hand, Rodenberg has also produced, with Ludwig Brill's **[225]** (now Martin Schilling's) publishing house a series of models of the F_3, connecting exactly with my studies.

If I have to start right away with the shape of the F_3, then the main principle is *that we take as starting point the figure already known with four real double points of p. [220] and construct from here more general surfaces through continuous transformation (untangling the double points) more general surfaces*[183].

We have already verified on the surface above, that now serves as an initial surface, the existence of the 27 straight lines, since namely three separate horizontal straight lines became added to the six edges of the tetrahedron, each of which is counted four times.

If we take as basis this surface's shape, then it is obvious that all the collinear transformations of the same surface are to be seen as equivalent, otherwise one

[183] See the corresponding procedure used for the plane algebraic curves of the n-th degree, where, by untangling, we passed from a curve with many double points, namely a curve composed by different curves of lower degree, to neighbouring curves.

could not have an overview of the manifold of the F_3 – whose equation contains 19 constants. This point of view is the same as when we do not distinguish, on the plane, between ellipses, parabolas and hyperbolas.

This is how it is theoretically. Practically, we are allowed to choose this somewhat noble standpoint only if we are experienced in interpreting collinear transformations. So we need, first of all, to know the different forms that arise from the initial model by means of collinearities. This can be done by intersecting the model with different planes, then projecting these as vanishing planes to infinity and studying the surfaces' forms thus obtained; and also making a model of them, in order to consider, afterwards, in all such shapes only the common properties, leaving the differences disregarded. *One can despise the knowledge of the details only when one dominates them, never before.* The result obtained is:

As well as we can distinguish ellipses and hyperbolas (here we omit parabolas, which represent the moment of transition) according to the relation of the conic section with the line at infinity, so we can distinguish, in the F_3 with four double points, five types according to their relation to the plane at infinity. For all these five types, their model has been constructed (see the note on p. [230]). We will no longer expressly distinguish among these five types.

From these surfaces, by untangling the double points, we pass over to the neighbouring surfaces.

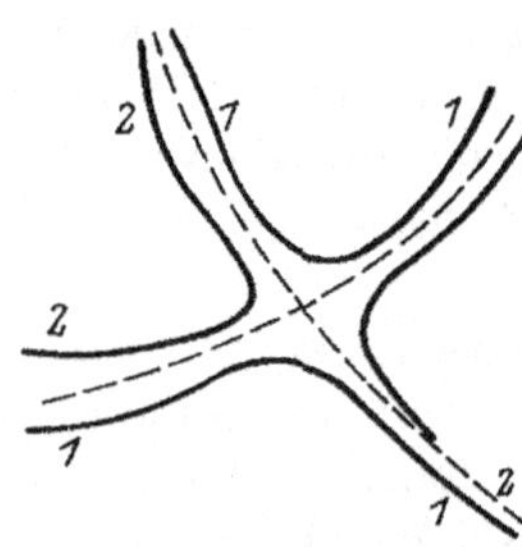

Figure 154

To this purpose, I first refer to the figure on the plane. We can clearly untangle
[226] a curve with a double point in two different ways for curves without double points (1 and 2 in Fig. 154). The two ways are per se equivalent. But if we go over to the three-dimensional case by rotating the curves around the vertical line passing through the double point, we obtain a significant difference in the arising surfaces. The first (1) is of the type of a two-sheet hyperboloid, which consists of two separate parts, the second (2) is of the type of a one-sheet hyperboloid, thus of a surface that presents – instead of the previous double point – a contraction (Fig. 155). We have therefore got the proposition: *A double point of a surface can undergo two different processes in the transition into a neighbouring surface, the process of*

separation and the process of merging. In the following, the process of merging will be symbolically denoted by $+$, that of separation by $-$.

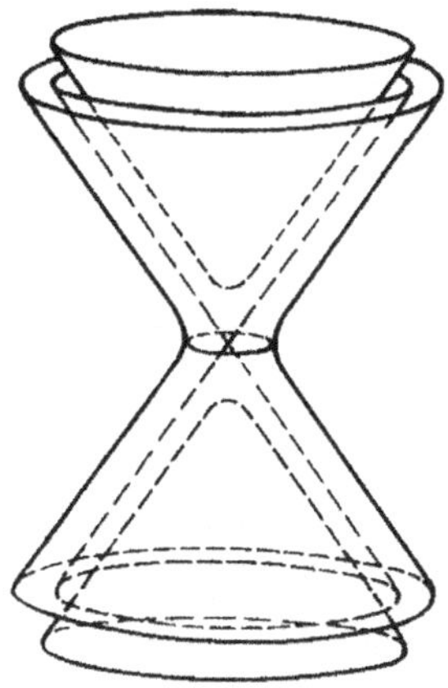

Figure 155

If we pass from our initial surface to the neighbouring surface in such a way that all the double points undergo both processes, then we have, summarising, five possibilities, indicated in the scheme

$$\begin{array}{lcccc} \text{I.} & + & + & + & + \\ \text{II.} & + & + & + & - \\ \text{III.} & + & + & - & - \\ \text{IV.} & + & - & - & - \\ \text{V.} & - & - & - & - \end{array}$$

Reflecting on the shape of these surfaces, we see that we obtain for the non-singular surfaces – due to our five possibilities – four one-sheet and a two-sheet surface ($- - --$). Indeed, in the latter case, by separation of all nodes, an ovaloid part will be surrounded freely by an undulated part of the surface; all remaining surfaces consist of just one part each.

These five cases correspond exactly to the types, which Schläfli has distinguished the F_3 according to the reality of the straight lines on them.

This results from the following consideration: An edge of the tetrahedron passing on the initial surface and connecting two double points is transformed, on the derived surface, into four imaginary straight lines, as soon as even *one* only of the double points, which is connected by the edge to another one, undergoes the process of separation. But on the other hand the following holds: An edge of a tetrahedron, which connects two double points, both of which undergoing the process of fusion, **[227]** is divided – in the transition to the neighbouring surface – into four real straight lines. According to this rule we obtain for our five surfaces

$$\begin{aligned} &\text{I.} \quad 3+6\cdot 4=27,\\ &\text{II.} \quad 3+3\cdot 4=15,\\ &\text{III.} \quad 3+1\cdot 4=\ \ 7,\\ &\text{IV.} \quad 3+0\cdot 4=\ \ 3,\\ &\text{V.} \quad 3+0\cdot 4=\ \ 3 \end{aligned}$$

real straight lines, where the last kind of surface differs from the preceding ones for the fact that it is made up of two parts. This corresponds exactly to the enumeration made by *Ludwig Schläfli*.

The relevance of this result is increased by the following proposition:

The representatives thus obtained of the five Schläfli surfaces free from double points are at the same time *fully valid representatives* of the individual kinds of surfaces *in the sense of the analysis situs*, that is two surfaces of the same Schläfli type emerge from each other by a simple continuous transformation of the shape, without the necessity of crossing a surface with double points; or even, all surfaces of the same type form a connected region in the representational[184] space of dimension 19 used for the representation – which presents us with the totality of the surfaces of degree three.

Thus you have, in the F_3 that we derive from the initial surface with four nodes using the processes + and –, typical examples of the five Schläfli types. Of course, it is important, for a concrete intuitive understanding, to transform the surfaces obtained by collinearity, or also to take the five types of surface of the F_3 with four real double points and to perform the processes + and – for each type.

Not only the F_3 without double points, but also the F_3 with arbitrary singular points can be obtained from our F_3 with four double points by means of a continuous transformation, and also in this case *Rodenberg*'s models yield numerous examples. Unfortunately, I cannot explain this more in depth here.

We ask furthermore: *How will the analytic study*[185] *of the models presented develop, and from which form of equation for* F_3 *is it better to start*?

This leads me to speak of a discovery of *James Joseph Sylvester*, namely that for the F_3 there exists one and only one so-called *pentahedron*, which is strictly related to the straight lines of the surface.

[228] For the investigation one needs to choose in a suitable way the five planes of this pentahedron as coordinate planes, that is so-called *pentahedral coordinates* are introduced. These coordinates are defined as following:

We consider the pentahedron formed by the five planes

$$x_i = a_i x + b_i y + c_i z + d_i = 0 \quad (i = 1, 2, 3, 4, 5),$$

---

[184] Translator's note: See note n. 163 on p. [185]

[185] Regarding the analytic study of the problem of the straight lines on the cubic surface, one can name among the more recent papers, *Bartel L. van der Waerden*, Der Multiplizitätsbegriff der algebraischen Geometrie, *Mathematische Annalen*, vol. 97 (1927), pp. 756–774.

no four of which should pass through a same point, and no three of which contain a same line. We choose as coordinates of an arbitrary point of the space its distances, multiplied by a fix constant, from the five pentahedral planes, that is we represent it by a system of values $x_1 : x_2 : x_3 : x_4 : x_5$. The five expressions x_i cannot be independent from one another, because a tetrahedron is already sufficient to determine homogeneous space coordinates. So the x_i represent so-called *supernumerary* point coordinates; the homogeneous linear relation among them can be expressed, with a suitable choice of the arbitrary constants, as

$$x_1 + x_2 + x_3 + x_4 + x_5 = 0.$$

Sylvester found that the chosen pentahedron of the F_3, named after him, yields such pentahedral coordinates, for which in the equation of the surface only the cubes of the x_i appear. The equation of the surface, consequently, is:

$$\sum_{i=1}^{i=5} a_i x_i^3, \quad \text{where it is} \quad \sum_{i=1}^{i=5} x_i = 0.$$

Taking this canonical equation of Sylvester as a starting point, *Rodenberg* proceeded to confirm the described geometric relations. The equation now contains only four of the 19 constants present in the original equation of the surface, and these four must necessarily remain, since only 15 can be eliminated by a suitable coordinate transformation. The planes $x_i = 0$ can here of course be real or imaginary.

We consider in particular the case of real pentahedron planes and obtain from Sylvester's equation a particularly simple F_3, that *Clebsch* called *diagonal surface*, by choosing the a_i all equal. This diagonal surface is a F_3 with 27 real straight lines; it is represented in our Fig. 156 more in detail.

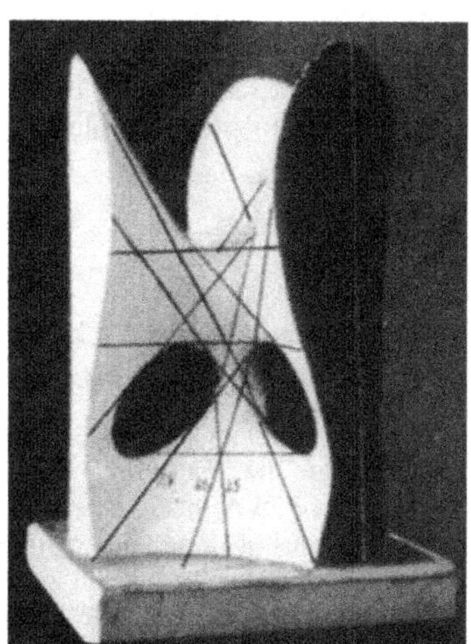

Figure 156

[229] The corresponding model was originally produced by *Adolf Weiler* at the direct request of *Alfred Clebsch*; *Weiler* was helpful also for my own research on surfaces of the third degree.

What is the particularity of this surface, where does its name come from?

If we intersect both the pentahedron and the surface with the pentahedral plane $x_1 = 0$, we obtain first of all, besides the plane $x_1 = 0$ itself, the quadrilateral built by the tracks of the remaining pentahedral planes $x_2 = x_3 = x_4 = x_5 = 0$, and then the curve given by the intersection with the surface

$$x_2^3 + x_3^3 + x_4^3 + x_5^3 = 0.$$

Now, *thanks to $x_2 + x_3 + x_4 + x_5 = 0$, it is possible to show that this intersection curve is composed of the three diagonals of the quadrilateral mentioned.* This explains the name of the surface. Therefore: *The name "diagonal surface" derives from the fact, that the surface contains – on each of the five pentahedral planes – the three diagonals of the quadrilateral obtained as section of the pentahedral plane with the four remaining pentahedral planes.*

Thus, at first the existence of $3 \cdot 5 = 15$ straight lines has been verified on the surface. I do not deepen the question of the 12 missing straight lines; you will easily find them on the surface model[186].

[186] [The existence of the 12 remaining straight lines can be seen analytically in this way: suppose $1, \varepsilon, \varepsilon^2, \varepsilon^3, \varepsilon^4$ are the fifth roots of the unity. Then it holds, as it is easy to see,

1) $$1 + \varepsilon + \varepsilon^2 + \varepsilon^3 + \varepsilon^4 = 0,$$

2) $$1^3 + \varepsilon^3 + (\varepsilon^2)^3 + (\varepsilon^3)^3 + (\varepsilon^4)^3 = 0.$$

It follows, that the system of values $1 : \varepsilon : \varepsilon^2 : \varepsilon^3 : \varepsilon^4$ yields a point lying on the diagonal surface. As we can arbitrarily interchange the five quantities $1, \varepsilon, \varepsilon^2, \varepsilon^3, \varepsilon^4$, without interfering with the validity of 1) and 2), we obtain $5! = 120$ total systems of values, which determine points on the diagonal surface. Among these 120 systems the ones that can be seen as distinct are only those, which do not arise from one another by multiplication with a fixed constant. If, for instance, we multiply $1 : \varepsilon : \varepsilon^2 : \varepsilon^3 : \varepsilon^4$ subsequently by $1, \varepsilon, \varepsilon^2, \varepsilon^3, \varepsilon^4$, and observe that it is $\varepsilon^5 = 1, \varepsilon^6 = \varepsilon, \varepsilon^7 = \varepsilon^2, \varepsilon^8 = \varepsilon^3$, then we obtain for the five equivalent systems of values, representing the same point:

$$\begin{aligned} x_1 : x_2 : x_3 : x_4 : x_5 &= 1 : \varepsilon : \varepsilon^2 : \varepsilon^3 : \varepsilon^4 \\ &= \varepsilon : \varepsilon^2 : \varepsilon^3 : \varepsilon^4 : 1 \\ &= \varepsilon^2 : \varepsilon^3 : \varepsilon^4 : 1 : \varepsilon \\ &= \varepsilon^3 : \varepsilon^4 : 1 : \varepsilon : \varepsilon^2 \\ &= \varepsilon^4 : 1 : \varepsilon : \varepsilon^2 : \varepsilon^3 \end{aligned}$$

Our 120 systems of values therefore represent only $4! = 24$ different points on the diagonal surface. We obtain them all, if we maintain the first term in $1 : \varepsilon : \varepsilon^2 : \varepsilon^3 : \varepsilon^4$ and carry out all the possible permutations on the others. These 24 points can be ordered in 12 complex conjugated pairs of points. The 12 real connecting lines of these pairs of points now completely lie on the surface. Indeed, if $x_1 : x_2 : x_3 : x_4 : x_5$ is the complex conjugate of one of our 24 systems of values $X_1 : X_2 : X_3 : X_4 : X_5$, a point lying on the line connecting the corresponding points is represented by

$$(\lambda x_1 + \mu X_1) : (\lambda x_2 + \mu X_2) : \ldots (\lambda x_5 + \mu X_5)$$

It is easy to verify that this system satisfies the equations 1) and 2) for each pair of values λ, μ].

Another proposition concerning the 15 straight lines, is that they intersect three by three ten times. One only has to consider that at each vertex of the pentahedron, representing the intersection of three planes of the pentahedron, three diagonals of the quadrilateral, one on each plane, converge and that the pentahedron has ten vertices. Therefore: *The 15 straight lines have (shortly) 10 intersection points, where 3 of the lines meet.* These intersection points are the vertices of the pentahedron. **[230]**

So I reached the end. You will now broadly understand the theory of the shape of the F_3; the details are reserved to your proper studies[187].

Appeal for an Always Renewed Adjustment of the Traditional Operating Mode of Science by Means of the Observation of Nature

Let me now conclude this lecture course with the following remark:

I have presented in these lectures various topics, which are commonly not found in the text-books devoted to the same themes, but which form the premise and the tacit assumption for the usual developments. With this, I aimed to induce you to grasp things by yourself, with open views and independent judgment. Just think of what I said about the empirical curve or surface, and of the otherwise usual restriction of the considerations to analytical structures.

---

[187] [A beautiful application of the fact that a F_3 contains 27 straight lines is proposed by *David Hilbert* in the paper: Über die Gleichung neunten Grades, *Mathematische Annalen*, vol. 97 (1927), pp. 243–250].

[Note to p. [225]. The five projectively equivalent transformations of the F_3 with four real nodes, mentioned on p. [225], can be derived from the surface of Fig. 149 in the following way:

The vanishing plane of the collinearity meets only the outer parts of the surface of Fig. 149, namely

1. in a one-sheet C_3 (without oval); the type is the same as shown in the figure;
2. in a two-sheet C_3 (with oval);

The vanishing plane also meets the tetrahedral form of the surface, and this again in a C_3 with oval:

3. the four nodes lie on one side of the vanishing plane;
4. three nodes lie on one side of the vanishing plane, the fourth on the other side;
5. two nodes lie on one side of the vanishing plane, the other two on the other side.

The occurring special cases are:

The vanishing plane is tangent plane of the

a) outer part: The intersection curve is a C_3 with ordinary double point;

b) tetrahedral part: The intersection curve is a C_3 with isolated double point.

The vanishing plane passes through

c) a node: The intersection curve is a C_3 with ordinary or isolated double point;

d) two nodes: The intersection curve is decomposed in a C_2 and a straight line;

e) three nodes: The intersection curve degenerates in three straight lines.

f) The vanishing plane contains the three simply counting straight lines of the surface.

g) The vanishing plane touches along the connecting line of two nodes.]

It is with mathematics as with fine arts. It is not only useful but also absolutely necessary to learn from one's forerunners. But if one confines oneself to the exclusive study of traditions, that is building only on what one reads in books, then what I call a scholastic system arises. Against this I am alerting you:

Go back to the own vivid conceptions, back to nature, which is always the best master!

Name Index

F. Klein, *Elementary Mathematics from a Higher Standpoint*,
DOI 10.1007/978-3-662-49439-4

Subject Index[1]

[1] Translator's note: This subject index, from the 1928 edition, often contains different terminology from the one used in the text.

D

Made in the USA
Monee, IL
13 September 2026

62973764R00155